Chemical Lake Restoration

Miltiadis G. Zamparas ·
Grigorios L. Kyriakopoulos
Editors

Chemical Lake Restoration

Technologies, Innovations and Economic Perspectives

Springer

Editors
Miltiadis G. Zamparas
School of Science and Technology
Hellenic Open University
Patras, Greece

Grigorios L. Kyriakopoulos
School of Electrical and Computer Engineering
National Technical University of Athens
Athens, Greece

ISBN 978-3-030-76379-4 ISBN 978-3-030-76380-0 (eBook)
https://doi.org/10.1007/978-3-030-76380-0

This Springer imprint is published by the registered company Springer Nature Switzerland AG
The registered company address is: Gewerbestrasse 11, 6330 Cham, Switzerland

Preface

Eutrophication has become the primary water quality issue for most of the freshwater and coastal marine ecosystems in the world. Eutrophication is considered one of the most visible examples of biosphere's alteration due to human activities affecting aquatic ecosystems from the Arctic to the Antarctic. As eutrophication becomes frequent and many eutrophic ecosystems face difficulties meeting the water quality criteria (EU Water Framework Directive), the removal of phosphate and/or ammonium gains great importance for water treatment.

Lake restoration efforts have traditionally focused on reducing nutrient inputs from the catchment, such as sewage discharges and diffuse runoff from agricultural land. Catchment remediation works—including restricting stock access to the lake, enhancing riparian buffer zones, and installing constructed wetlands on the main inflow—they have been previously used. However, there are globally observed highly enough in-lake concentrations to maintain eutrophic conditions. Thus, even in cases of successful remediation treatments the recovery of the waterbody may still be at a slow pace due to excessive internal loads of nutrients.

Phosphorus release from the sediment into the lake water may be so intense and persistent that it prevents any improvement of water quality for a considerable period after the loading reduction. External nutrient loading over time increases the concentration of nutrients in the sediment that can be released back into the overlying water column. After receiving elevated external inputs of P for decades wetland systems are greatly impacted by internal P flux, a phenomenon referred to as "Phosphorus legacy". The book provides a cross-sectoral, multi-scale, and sequential way upon the development of knowledge related to eutrophication control. The book's content contains the latest developments of innovative technologies for phosphate removal while supporting the restoration of currently degraded lakes and reservoir systems. Subsequently, the book unveils innovative approaches in the light of benchmark opportunities offered and circular economy principles applied.

The uniqueness of the book resides to the fact of presentation and co-evaluation of innovative approaches to phosphate removal, supporting the inland water restoration at an integrated management. The book content is not limited to the technological viewpoints and overview but it is extended to the economic appreciation of such niche technologies, including the role of circular economy at the marketable perspectives offered. Consequently, the book offers an integrated approach of better understanding the problem of internal P-loads in the light of effective controlling nutrient enrichment in lakes, according to circular economy considerations. Since the book's thematic is extended to economic appreciation of environmental polluted aquifers, the book audience is covering a wide spectrum of scientific fields, such as environment, physical chemistry, aquatic chemistry, coastal engineering, bio-engineering, and environmental policymakers. Moreover, P-adsorbents, as materials for fertilizers, can be appealing to soil and plant science, including agriculture and ecology. The book is also attractive for researchers of water science, including marine ecologists, material scientists, and chemical engineers. Linkages developed between economic aspects and technological applications can be of utmost importance for satisfactory environmental management and landscape design, making it an attractive and pluralistic must-read book for academic purposes among university students and academic teachers, as well as among researchers and policymakers of environmental planning and water sources' remediation.

We are grateful to all the contributors of this book for their intellectual researches, insightful propositions, and creative arguments. We also want to warmly thank the professional staff at Springer Nature editions for their qualitative work that made this book possible. We hope you to find the contents of the book fascinating and its reading attractive. If you would like any further information on this book, we are always at your disposal.

Patras, Greece — Dr. Miltiadis G. Zamparas
Athens, Greece — Dr. Grigorios L. Kyriakopoulos

Contents

Chapter 1
Eutrophication—A Worldwide Water Quality Issue

Vasileios C. Kapsalis and Ioannis K. Kalavrouziotis

Abstract In this introductory chapter of the book, the description of the eutrophication phenomenon was developed, while the generating causes of the phenomenon and its effects on the natural environment have been investigated. Besides, in this chapter the research has focused on the phosphorus cycle in water, disclosing the critical contribution of internal nutrient loads to the eutrophication phenomenon. The contribution of internal nutrient fluxes has been systematically approached in alignment with several examples of ecosystems' suffering from eutrophication, while the managerial need of these nutrient sources to mitigate eutrophication has been signified. Finally, the socio-economic effects of eutrophication were reported, showing that eutrophication intensity in water sources is a deterring phenomenon of the recreational activities that were also resulting in a loss of income. There are not precise and consistent norms of valuating financial losses while failing to apply suitable eutrophication treatment at natural water sources (for others, being appreciated as invaluable). However, in a pragmatic scientific approach, the expected per-person annual damages should range at tenths up to hundreds of €, while the communal collective responsibility among densely inhabited populations can reach tenths of billion €, or even more. Therefore, the significance of appreciating natural water sources and keeping a good eutrophication status remains a critical, yet disputable and contentious, issue.

V. C. Kapsalis (✉)
Sector of Industrial Management and Operational Research, School of Mechanical Engineering, National Technical University of Athens, 9 Heroon Polytechniou Street, 15780 Athens, Greece
e-mail: bkapsal@mail.ntua.gr

I. K. Kalavrouziotis
Laboratory of Sustainable Waste Technology Management, School of Science and Technology, Hellenic Open University, Parodos Aristotelous 18, 26335 Patras, Greece

M. G. Zamparas and G. L. Kyriakopoulos (eds.), *Chemical Lake Restoration*,
https://doi.org/10.1007/978-3-030-76380-0_1

Graphical Abstract

Graphical Abstract: Eutrophication as a worldwide water quality issue. A word cloud

Keywords Eutrophication · Phosphorus cycling · Water quality · Algal blooms

1.1 Eutrophication—In General

Eutrophication is the natural phenomenon observed when phytoplankton blooms are caused by excessive amounts of nutrients (mainly N and P) in water [1–3]. Today the rapid increase of global population and the improvements of well-being at humanity are depleting natural sources thus resulting in ecosystem endangering, difficulties of control, and slow restoration, especially in the field of preserving the quality of water sources and confronting the ongoing problem of global water pollution [4–6]. The Ministry of Environmental Protection, PRC, in China published the "2008 Environmental Condition Bulletin of China" that contains data on various water sources in China. It was reported that among 26 lakes and reservoirs, only 1 lake met Class II water quality standard (counting for 3.9%), 5 lakes met Class III (counting for 19.2%), 6 lakes met Class IV (counting for 23.1%), 5 lakes met Class V (counting for 19.2%), and 9 lakes met Class V (counting for 34.6%). The quality order of such a classification follows the ascending ranking: Class I (lowest) up to Class V (highest). Regarding the nutritional status reported and the 26 lakes and reservoirs monitored, one lake was severely eutrophicated (counting for 3.8%), two lakes were moderate eutrophicated (counting for 7.7%), eight lakes were mildly eutrophicated (counting for 30.8%), and the rest lakes were mesotrophic [7].

The ongoing human activity is worsening eutrophication, caused by changes in geochemical cycles of carbon (C), nitrogen (N), and phosphorous (P). In particular, these elements' balancing out is affecting the inflow of nutrients to aquatic ecosystems through point and non-point sources of anthropogenic activities of various aspects of community: municipal, industrial, agricultural, manure and catchments. Non-point sources are larger, of higher importance, and more complex to be controlled, comparing to that of point sources [1, 8–10] (Fig. 1.1).

Five decades ago, the structure and the causalities of eutrophication were not well known or fully understood, even though a plethora of scientific production has been devoted to these issues. Nutrients were preliminary the suspected determinants of

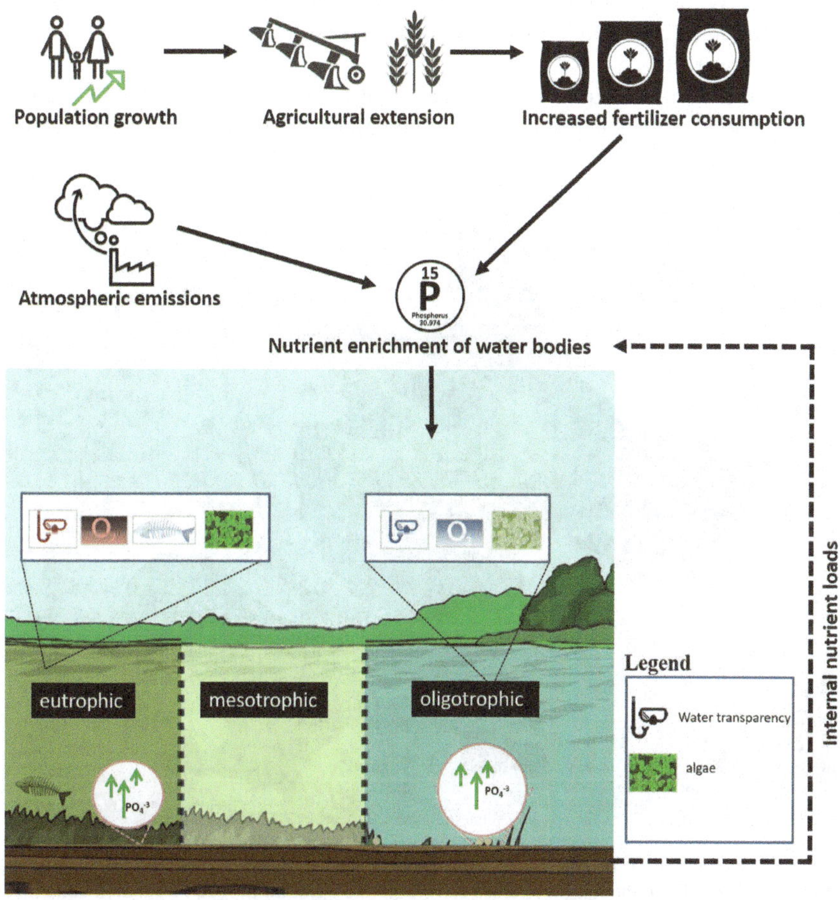

Fig. 1.1 Drawing of the eutrophication process determined by the synergistic action of direct and indirect contributors. *Source* Authors' own study

eutrophication, but there was strong evidence on them. Moreover, the algal abundance was associated with high amounts of many dissolved compounds, mainly including N and P [11, 12].

Eutrophication could be easier correlated with the concentration of inorganic carbon under experimentation of aqueous samples at closed vessels. Moreover, at other container experiments, it has been reported that since N and P are the fundamental feed of algae blooms these two elements can satisfactorily determine the growth of algae [13]. Among other factors, physiological indicators have been indicated that N and P were evenly limiting to algae, apart from for algal blooms where cells are indicating N shortage. The N to P ratio is a feasible indicator of algal cells, implying Nitrogen limitation through algae blooms, as it was signified after whole-lake studies at Canada's Experimental Lakes Area [13, 14]. These studies clearly argued that eutrophication is exclusively caused by P generated by the production of human interventions causing massive growth of phytoplankton, whereas by enriching the reference systems with inorganic N and C, it was shown no pronounced changes [13, 14] (Fig. 1.2).

Among P-rich lakes quantified indicators of C and N at algae blooms are that of inorganic C diffusion from the atmosphere and N_2 fixation by cyanobacteria.

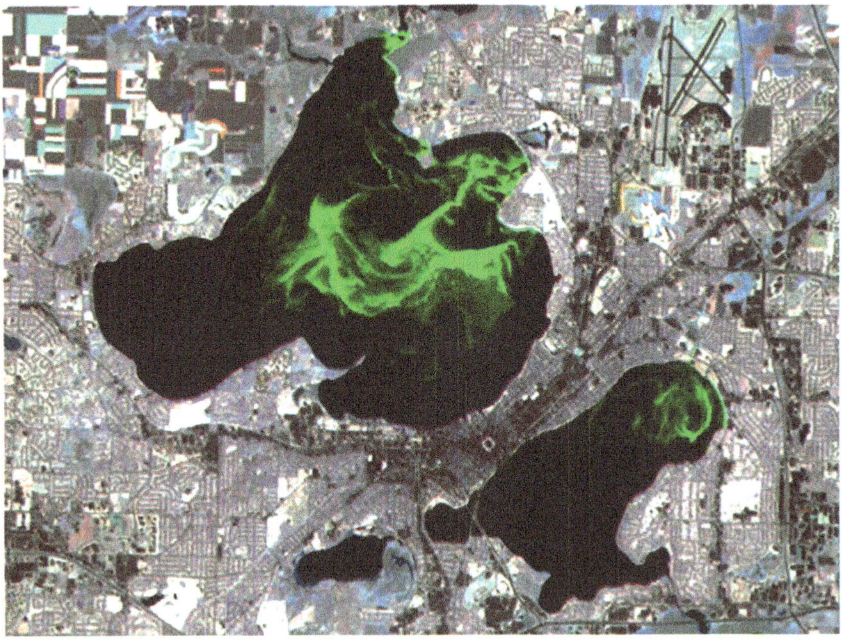

Fig. 1.2 Cyanobacteria blooms (*Microcystis aeruginosa*) in lakes Mendota and Monona, Wisconsin, USA. Even though these lakes can temporarily exhibit nitrogen limitation conditions during summer blooms, lakes' eutrophication is driven by P runoff from agricultural lands and urban areas. In this context, the false-color *LandSat* image, above, processed to highlight surface bloom [13]

However, the generalized causality of N limitation and the concurring reduction of algae blooms cannot be evidenced from a literature point of view, since knowledge is mainly coming from artifact experiments of closed containers and short-time sessions, whereas eutrophication of water sources is actually a prolonged and slowly evolving natural phenomenon [15].

Based on extensive and last-longing literature production, as it has been also presented above, P has proven of pronounced responsibility at natural water sources' eutrophication, as well as uncontrolled aquatic plant and algal growth [16, 17]. Such excessive, mainly anthropogenic-caused, growth of aquatic plants is accompanied by decomposition of organic matter, low concentrations of dissolved oxygen in hypolimnetic waters and sediments, low water renewal rates [18], as well as nurturing of hypoxic/anoxic conditions among global waters, in stratified rivers, lakes, reservoirs, estuaries, and coastal waters [19–21].

1.2 Lake Classification

To successfully assess lake restoration technologies, a study of a waterbody's morphological and physicochemical parameters is necessary. Lakes are normally categorized into four types; oligotrophic, mesotrophic, eutrophic, and/or hypereutrophic. Low productive oligotrophic water bodies are typically clear, with low P and N concentrations and few macrophytes (P concentrations <1 μg/L) [22]. These lakes maintain adequate dissolved oxygen (DO) in the bottom layer during late summer. Mesotrophic lakes are in an intermediate state, while eutrophic lakes are characterized by high nutrient concentrations. High-productive eutrophic lakes are normally shallow, turbid, with rich aquatic plant growth. In deep lakes, the hypolimnetic waters typically hold small concentrations of dissolved oxygen. Moreover, hypereutrophic lakes are those lakes characterized by very high productivity, nuisance algae, and weed growth. Such a lakes' classification referring to trophic states, it has been determined by Secchi depth, Chl-a, and TP and denoted in Fig. 1.3 [23].

1.3 Lakes Versus Ponds

Nowadays, there is an increasing interest in the eutrophication process and restoration techniques of ponds. Ponds have been mostly neglected in research regarding freshwater systems, and only in the last decade, they have begun to be documented and examined as unique ecosystems, separate from lakes, streams, and rivers [25, 26]. This is significant to distinguish in eutrophication assessment because aquatic ecosystems will experience eutrophication differently [27]. There most widely accepted definition of ponds comparing to lakes definition, which has been based on size. Consequently, Biggs et al. [28] set the site-criterion in between 1 m^2 and 2 ha while Oertli et al. [29] added the depth-criterion of no more than 8 m. A shallow depth can

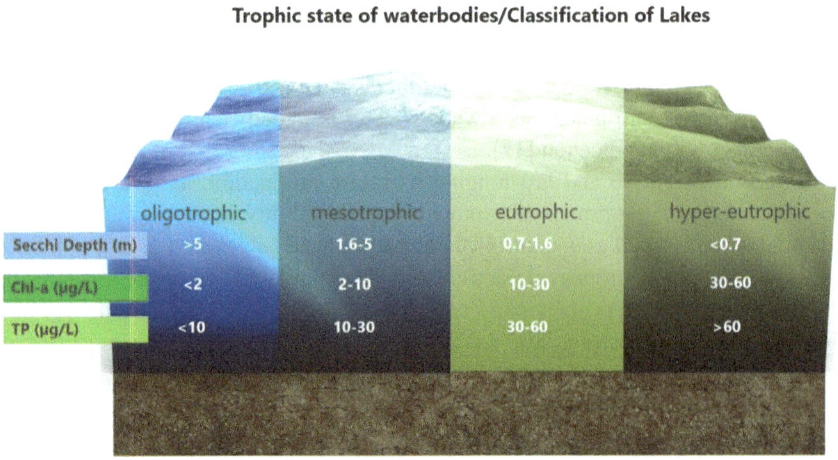

Fig. 1.3 Classification of lakes. *Source* Modified from [24]

keep stable temperature throughout the water column, being dependent on changes in air temperature. A shallow depth is potentially responsible for the growth of aquatic vegetation, also typically extended to shorelines covered by vegetation. The shallow ponds are also able to offer longevity of residence time than lakes, sustaining a longer residence time than deep lakes. Regarding dissolved oxygen, it is highly available during the daytime and gradually depleted during the night. A depiction of biomass pyramid in ponds can be graphed in a large base for plant-life, about 87%, followed by herbivore and carnivore masses of 10 and 3%, respectively. However, endemic communities in ponds are changing only in a dynamic seasonal and daily variability, as a result of variability in the shallow water column. A distinction between natural and man-made ponds is not seriously significant, while the distinct characteristics of the pond, compared to lakes, are referred also to ecology differences from each other. In an infield study comparing lakes and ponds, researchers denoted that larger water bodies can support greater biodiversity [25, 26, 30].

In a study specifically measuring differences in biodiversity between ponds and lakes, Hamerlik et al. [31] argued that water bodies of two hectares, the maximum, are showing distinctly different species–area relationships than larger water sources. Lakes sustain a significant positive correlation between size and diversity, whereas pond diversity is less bounded on size. However, in the light of areas covered by ponds and lakes, it is noteworthy that ponds sustain higher diversity than lakes, as well as thriving aquatic biodiversity. Biodiversity could be also associated with the fact that ponds, having a smaller size than lakes, are prone to more variety in their physical and chemical makeup than lakes. This fact is implying nurturing specific endemic species composition and eutrophication routes that are processed are different between ponds and lakes since these nurture conditions and processes are largely driven by varied physical and chemical aspects. Therefore, accumulated knowledge has, yet, to be developed to better understand the exact functionality of pond-vs-lakes processes, being seemingly unlikely that the application of the same ecological standards for lakes and ponds would be equally satisfactory for eutrophication management [31, 32].

1.4 Phosphorus Cycling in Sediments

In most aquatic environments, P is deposited into the sediments from the overlying water column mainly in the form of organic P, as part of settling organic matter [27, 33]. In contrast to the marine environments, the P content of freshwater organic material is significantly more variable. Between 20 and 90% of the organic matter deposited to the sediment surface is typically mineralized within the sediment, and the fate of the mobilized soluble phosphorus depends on whether it is re-sequestered within the sediment or diffuses upward into the water column. Mineral phosphorus can be deposited to the sediment surface as apatite-P or as P adsorbed to calcite, iron, or aluminum minerals or clays [34–36].

The sediment geochemical cycling of P is tightly coupled to the cycling and redox dynamics of Fe (Fig. 1.4). When the water column overlying the sediment contains oxygen, the oxidizing conditions at the sediment surface favor the precipitation and the preservation of ferric (Fe(III)) particles in the oxidized upper layer of the sediment. Fe(III) phases include iron oxides and, importantly, oxyhydroxides (ferrrihydrite, goethite, lepidocrosite), being denominated as nanophases of large

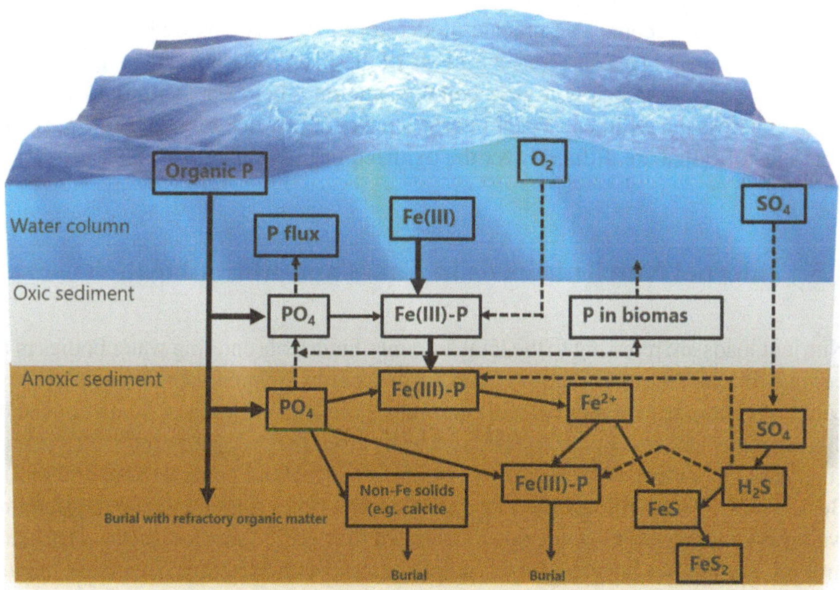

Fig. 1.4 Simplified diagram of P cycling in sediments. *Source* Modified from [27]

specific surface areas, high bioavailability, and a strong capacity for binding P. The high insolubility of ferric iron was reported under oxic conditions and circumneutral pH, while their burial into the deeper anoxic sediment is undergone microbial mediation and reduction to Fe(II). A plethora of microbial species (such as that of Geobacter and Shewanella genera) are prone to catalyze this reaction, coupling it to the oxidation of organic carbon in their metabolism types: dissimilatory or sometimes fermentative:

$$4Fe^{2+} + O_2 + 10H_2O \rightarrow 4Fe(OH_3) + 8H^+.$$

The aforementioned process enables iron mobility, thus, releasing it as soluble Fe^{2+} along with Fe-bound phosphorus, into sediment porewaters (Fig. 1.4). At the majority of sediments the oxygen penetrates it by a few millimeters and the proximate underlying zone of oxygen penetration is a site of intense redox cycling. The release of dissolved Fe^{2+} at anoxic sediments can diffuse upward, where it can oxidatively reprecipitate near the oxic–anoxic boundary, form Fe(III) oxyhydroxides again, and bind porewater phosphate [37]:

$$2Fe^{2+} + MnO_2 + 4H_2O \rightarrow 2Fe(OH)_3 + Mn^{2+} + 2H^+.$$

Continued cycling and recycling of Fe across the sediment redox boundary, on or slightly underneath the depth of oxygen penetration, can result in the accumulation of Fe, while producing typical Fe-rich layers of several millimeters thickness [38].

Fe-rich layers are serving as a barrier to the diffusion of phosphate, which is produced upon the decomposition of organic matter in deeper sediment. Therefore, redox dynamics become a significant moderator of P efflux from sediments. Within the oxic sediment layer, the concentrations of dissolved phosphate are significantly lower than in the anoxic porewaters below (sometimes below detection). Besides, P effluxes across the sediment–water interface are certainly determined by the vertical gradients of P that are diffused over the oxidized near-interface zone [27, 39].

1.5 Internal Nutrient Loads in Lakes and Eutrophication

Nutrient loads are referring to the total amount of nutrients entering water bodies in a definite time and are one of the key determinants of phytoplankton biomass. Nutrients content is composite of internal and external sources [40–42]. It is indicatively noted that nutrients in lakes and reservoirs are mainly from the exogenous inflows (that are considered as external) and endogenous outflows (that are considered as internal). Besides, exogenous inflows are referring to natural flows from rainfall and upstream water, as well as anthropogenic flows generated from varied sources, such as fertilizer inflows and inflows generated from point sources, such as sewage treatment plants [43–45]. As endogenous inflows are considered the release of nutrients from the dumped soil, primarily related to redox reactions at the sediment–water interface and the subsequent decomposition of the settling organic matter [46]. The main types of upstream water quality of reservoirs are linked with the industrial structure and the spatio-habitual uses of the basin as a receptor of industrial production, the loss of fertilizers, rural residency outflows, livestock and poultry breeding, land erosion, and fishery [47]. Contaminants from the upstream basin are either directly or indirectly discharged into the reservoir having a pronounced role in the reservoir water quality [48, 49].

Besides, long-term exogenous nutrient inflows and aquatic organisms' residue depositions are leading to the accumulation of excessive quantities of nutrients from sediments, which are easily released by molecular diffusion, since this phenomenon is attributed to differences in concentration between nutrients in the sediment–water interface [50]. Nutrients' release is also determined by a range of parameters, such as wind, benthic organisms, bubbles overflow, phytoplankton floating, and transferring through aquatic plants and their roots. This multifaceted phenomenon is determining the increase of nutrients' content in water sources, thus, promoting algal blooms [51–53]. Nutrients' sedimentation at natural waters is certainly resulting from eutrophication occurrence, especially among lakes' and reservoirs' proximity at household and domesticated areas [53, 54].

Phosphorus concentration is valued as an important and measurable indicator of the productivity levels and the eutrophication episodes at inland waters. In natural lakes, the main phase of P-existence is in organic-P. Moreover, a 70% portion of organic phosphorus is particulate matter, thus, it supports P settling at the bottom of the water column after entering water [55]. Subsequently, sentimental concentrations of P are much higher, comparing to those in the overlying water phase. Under certain conditions P-release into the overlying water is observed, causing P to be relocated and considered as endogenous load. Migration of P from the sediment into the overlying water can be reported as a persistent phenomenon for prolonged periods, thus, delaying the water quality improvement at a considerable time after the occurrence of external loading (Fig. 1.5) [55]. Similarly, Søndergaard et al. [56] reasoned the aforementioned slow response to decreased external loads due to storing of P-sediment can function as a buffer to P concentration alterations at water column [56].

Kagalou et al. [57] denoted that the instantaneous reduction of external load results in a C_{lake}/C_{in} ratio increase in the water column, where C_{lake} counts for the internal concentration of nutrient concentration and C_{in} counts for the inflow concentration of the nutrient. Then, C_{lake}/C_{in} decreasing ratio is expected thereafter, when the P content in the top sediment layer has reached equilibrium with the elemental concentrations in the overlying water column. However, the recovery is a time-demanding process that varies considerably from lake to lake, necessitating decades in some cases [57, 58].

Another insightful remark is that over time the sentimental nutrients' concentration should be released into the overlying water column by increasing the external nutrient loading [40, 59]. In such an overtime natural process of accruing P-receipt of elevated external inflows, there are wetland systems that are greatly impacted by internal P-flux (e.g. Florida Everglades), the so-named "phosphorus legacy" phenomenon [8, 60] (Fig. 1.5b). The sediment plays a decisive role in the profile and

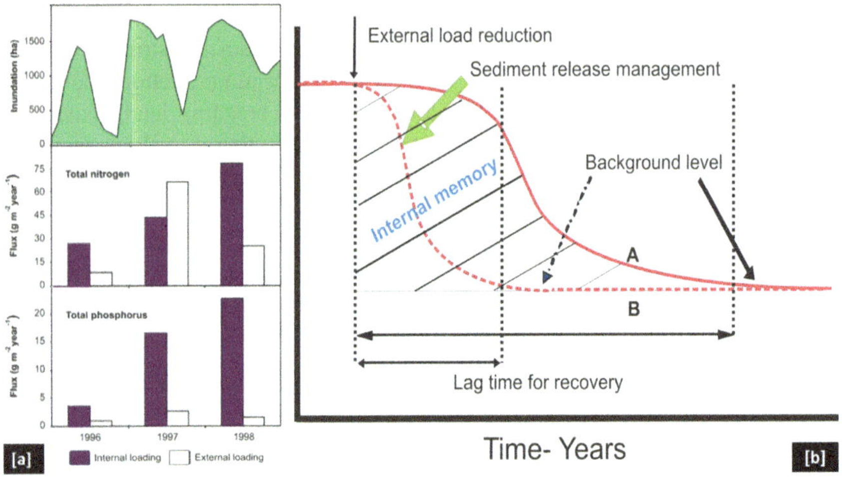

Fig. 1.5 **a** External and internal annual loads of P and N and inundation in Las Tablas de Daimiel from 1996 to 1998, **b** Line A (solid line)—refers to the correlation between water P column and internal P memory due to legacy phosphorus. Line B (dashed line)—refers to the correlation between water P column, internal P-memory, and implementation of sediment release management. *Source* Modified from [7, 8]

the dynamics of nutrients at shallow lakes. Kagalou et al. [57] focused their study on a restoration plan at the Lake Pamvotis (Greece) that involved sewage diversion and led to the reduction in external nutrient load, while lowering the concentration values for the in-lake nutrients and chlorophyll-a. Regarding other co-existent compounds, it was reported that orthophosphate concentration decreased by about 87%, nitrates dumped below 1.20 mg/L, whereas the total reduction of inorganic N compounds expressed a retarding plateau-downward behavior, ranged at 0.39–1.24 mg N/L, that is 0.76 mg N/L in average. While the aforementioned restoration was undertaken from 1995 through 1996, after a short-term recovery, it was reported persistence of the lake Pamvotis eutrophication up to 8 years later (1996–2004). This observation unveiled the utmost important role in the internal loading process. Indeed, it was expected that internal loading can enhance eutrophication and it can decrease the natural self-remediation of lakes' response-time, in response to the external reduction in P load, by a time-frame of a decade, or longer [57].

Several similar studies demonstrated that the internal loading of P in the nutrient dynamics of wetlands is affected by long-term nutrient loading [7, 61]. In such a case, the Las Tablas de Daimiel semiarid fluctuating lake of Central Spain showed that internal loading is the most prevailing P-source to this ecosystem, even in cases that high external inputs occurred through time (Fig. 1.5a).

Moreover, at another long-term data-based study regarding 35 lakes in Europe and North America, it was observed that internal P release typically occurs for more than 10–15 years after external loads' reduction [55], and in some lakes, P release from the sediments last longer than 20 years [55]. According to Jeppesen [62] in lakes where the external loading has been reduced, internal P release may avert improvements in the quality of freshwaters. At extreme internal loading, particularly summer concentrations rise, and phosphorus retention can be negative throughout the summer period [62]. Consequently, remediation of sediments must often be carried out to enhance lake recovery.

Tuusulanjärvi in 1979 concluded that the decrease of external P loading did not contribute to water quality improvement during the following three decades. Moreover, internal loading of P showed an increasing behavior, then decreasing P-anticipated, during the coming 2000s. Here it is important to recall that large-scale artificial aeration efforts put into action in the late 1990s and they affected the P dynamics, but these effects have not been unambiguously positive. In particular, a decrease of anaerobic P-release from the sediment was reported, the total internal P loading increased. This P-flux increase can be probably attributed to the destratification of the water column both artificially and in severe changes in weather conditions [63].

P entering the sediment is followed by numerous processes of chemical and biological mediation, enabling to be: (a) ultimately or permanently deposited in the sediment or, (b) undergone several release-processes and return in dissolving form to the water column via the interstitial water [64]. Thriving literature studies have been devoted to introducing mechanisms proposals for P-release from lake sediments. However, it is important to stress out that quantitative and qualitative, mainly chemical composition, characteristics among lake sediments can be substantially varied [65]. Typical parameters having the capacity of P-binding and P-release are influenced by the sediment–water interaction aspects, including dry to weight ratio, organic content, as well as quantifying elemental content of iron, aluminum, manganese, calcium, clay [62, 65–68].

The main detrimental phenomena concurring to high internal P-loading are that of concentrations rise and the disadvantageous P-retention. In particular, the internal P loading intakes from a pool accumulated in the sediment at high external loading, while significant P quantities in lake sediments are prone to be either, (a) bound to redox-sensitive iron compounds or (b) fixed in more or less labile organic forms. These are commonly mobile forms that can be eventually released into the supernatant lake water [69, 70]. Among those factors implicated in sedimental P-release to the water column, they are the prolonged external loading, thus, ultimately producing thick levels of nutrients' sediment, rebounding them to water column under certain environmental conditions, for example, anoxia [71]. Therefore, persistent high nutrient concentrations of water column have been reported at the considerable plethora of lakes, regardless of external loads' reduction occurred, especially from point source inflows. It is noteworthy that research studies of internal P loading are playing a vital role for researchers to better understand the prevailing processes of P circulation and, subsequently, to support feasible policies in eutrophication management [40, 57, 72–75].

In valuing the notable role of phosphorus (P) in the eutrophication process, one successful restoration method particularly among shallow lakes is the removal of sediments enriched with nutrients (Fig. 1.6). In a recent study [76], annual, seasonal,

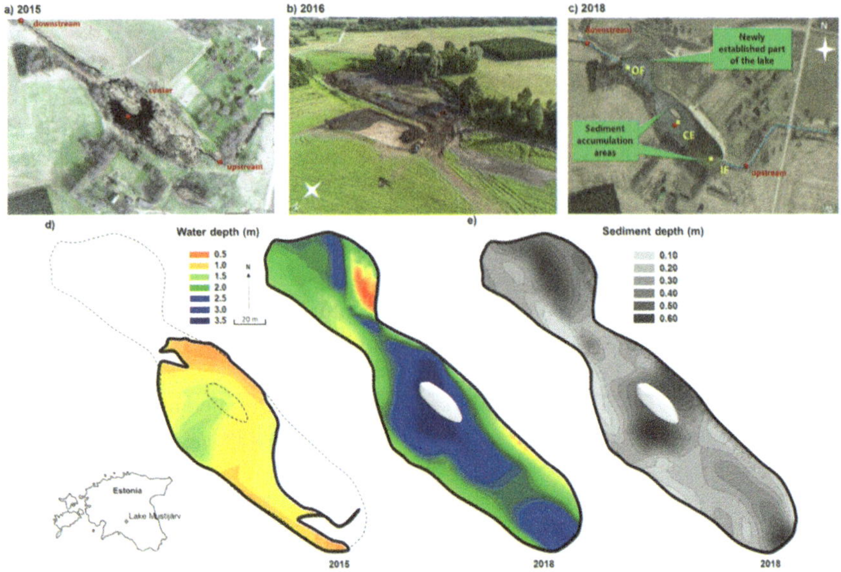

Fig. 1.6 **a** Photo of eutrophic Lake Mustijärv in Viljandi (Estonia) and sampling locations in 2015, **b** sediment removal of the lake in 2016, **c** sampling locations in 2017 and 2018 (yellow points), **d** water depth in 2015 and 2018, **e** depth of newly formed sediment measured in 2018. *Source* Adapted from [76] with permission from Elsevier

and spatial data were devoted to the P-mobility changes' examined after the removal of sediments from a one-hectare highly eutrophic lake, the year 2016. The authors investigated the redox potential (Eh) of sediment, measured soluble reactive phosphorus (SRP) in the sediment–water interface and P-fractions of the surface sediments and, then, calculated the P-diffusive flux at three sampling sites in the 2017 and 2018 subsequent years after the sediment removal. Similarly, measurements were made in the preceding year 2015, prior to excavation at the central location of the lake. Nutrient-rich sediment also removed 6400 kg of P, while at a time running the long-term release of sedimental P is decreasing, accordingly. However, large quantities of P-releasable were quickly rebuilt after the sediment removal, mainly due to high external P loading, which caused extensive anoxia at the sediment surface in alignment with internal P loading reached at high 1450 mg m^{-2} $summer^{-1}$. Besides, Fe–P and labile-P fractions were prevailing sources of P release, as reported by sediment removal changes, either seasonally or inter-annually referred. The sentimental Fe-total was reversely related with diffusive P-flux, pore water SRP, and near-bottom water total P and SRP concentrations proving the development of close associations between (sediment P dynamics) and (Fe after the restoration). In such an approach, sediment removal could benefit the restoration approach only in the short-term, unless the effects on lake water quality are adequately controlled by external to-the-lake loads [76].

Whenever internal nutrient loads have to be decreased for macrophyte control, or removal of toxic substances present in the lakebed is needed, sediment removal is suggested [77]. However, the key aspects of dredging works are affected by practical considerations including the determination of how much sediment to be removed, determinate the collective site of disposing the after removal of contaminated sediment, the techno-economic feasibility of relocating sedimental matter, as well as the geo-morphological distortion endangered, being the consequence of contaminants' release into lake water at dredging operation [78]. It is exemplary noted that in Alderfen Broad, UK, although the exposed sediment sustained lower P quantities than that of P-removed sediment within a year, the total sedimental P quantity was increased by 80% of its original value. Therefore, P concentration continued higher than that of prior-to-eutrophication occurrence [79].

1.6 Harmful Algal Blooms and Eutrophication

A serious adverse effect that is linked to eutrophication, having significantly impacting on public health and economy is the presence of harmful algal blooms (HABs) in natural eutrophic water bodies [80–82]. HABs are potential threats to human and aquatic ecosystem health. Besides, economic losses that are associated with HABs, possibly involve loss of recreational revenue, depreciation of property values, and increasing costs of drinking-water treatment [83].

Algal blooms of numerous nontoxic species are harmful through their high biomass content, likewise to toxic species, forming foams or scums on the surface water, depleting oxygen as blooms decay, destructing flora and fauna accommodated at aquatic and submerged habitats [84, 85]. Among the most harmful algae since the beginning of the second half of the twentieth century is that of Cyanobacteria blooms in freshwater basins while causing wide detrimental consequences of nuisance or other environmental depletion issues, including thick hyperscum mats, production of potent hepatotoxins and neurotoxins, causing livestock and wildlife death, and occasionally human death [86, 87].

At a historical flashback report of eutrophication through the period 1960s and 1970s in the Seto Inland Sea (Japan), it led to a sharp increase in red tides incidents by (indicative): *Raphidophyceae* and *Dinophyceae polykrikoides* species [88–90]. In Japan, the effect of HABs on living resources was recorded at differentiated coastal and marine ecosystems, where various toxic species resulting in detrimental effects on the fish stock, causing mass mortality of variable aquatic animals and generated unrecoverable economic damage [91–93]. In Europe, the main consequences of exposing various algal blooms, by the toxic dinoflagellate species [94, 95], are that of water discoloration [94], undesirable esthetic symptoms [96], anoxic detrimental conditions to the aquatic ecosystem [97], and fish mortality events [98].

In the United States, toxic freshwater cyanobacterial algal blooms were condemned as leading factors of human and animal illnesses and increased deaths among at least 43 states. In August 2016, 19 states developed public health advisories because of cyanoHABs (Fig. 1.7) [99].

Despite advances in scientific knowledge about cyanobacteria, many unanswered issues remain regarding the incidence, environmental triggers for toxicity, and the ability to predict the timing, duration, and toxicity of CyanoHABs [100]. Scientific data and mechanistic understanding of environmental factors—as well as the related adverse effects of cyanotoxin exposure—are vital to building credible warning systems and prognostic tools that play an important role in decision making [99, 100]. Advanced warning at time scales relevant to CyanoHAB management in water bodies throughout the United States, including the following: (a) developing laboratory and field methods to detect and calculate CyanoHAB compounds, (b) understanding causal factors, environmental fate and transport, ecological processes, and effects

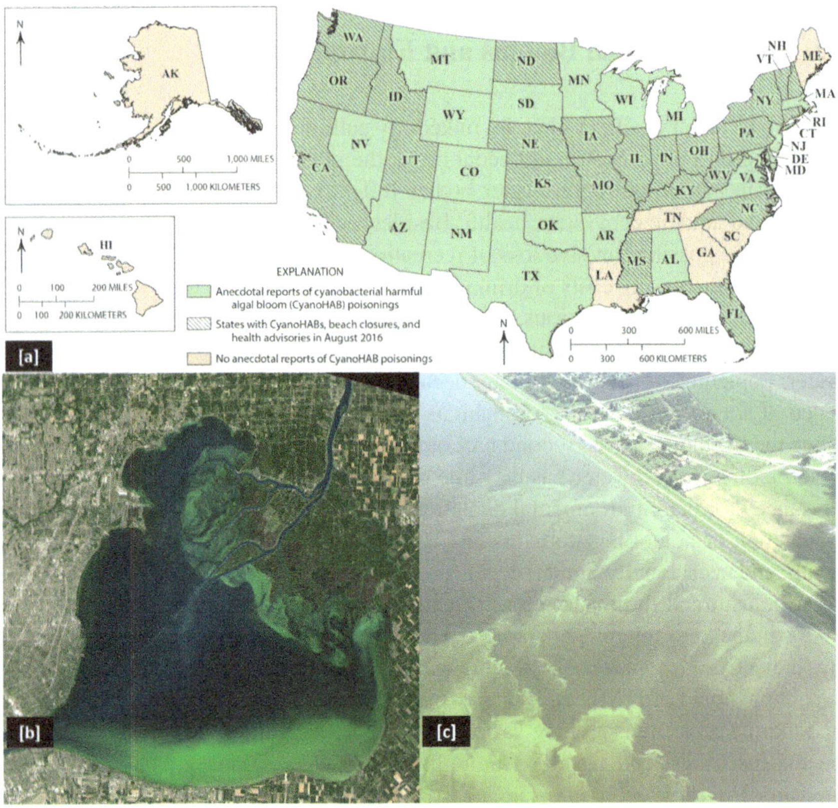

Fig. 1.7 **a** Toxic cyanobacterial algal blooms have been involved in human and animal illness and death in at least 43 states in the United States, **b** aerial view of cyanobacterial bloom, July 2016, Lake Okeechobee, Florida, **c** Satellite view of a harmful cyanobacterial algal bloom, Lake St. Clair, U.S. State of Michigan. *Source* Modified from [99]

of environmental exposure, and (c) developing early warning systems for potentially harmful blooms, including analytical techniques, stable isotopes, molecular techniques, sensor technology, and satellite imagery [100].

1.7 The Socio-economic Aspect of Eutrophication

The economic aspects of eutrophication at natural waters are linked to recreational angling and boating activities since these activities can be directly and physically prevented by eutrophication-driven macrophyte growth and algal blooms [101]. Moreover, water leisure activities are made complex and dangerous, like swim, boat, and fish during heavy algal blooms, mainly regarding health risks, unpleasant appearance, or bad odors. Property values can be also depreciated by lower water transparency/visualization [101]. It is also a noteworthy fact that the potential annual value losses, due to alteration of habitual uses of recreational water and waterfront real estate, can be approximated to $2.2 billion per annum as a consequence of lake eutrophication in the United States [102]. The main portion of such economic losses can be ascribed to lakefront property values ($0.3–2.8 billion per year) and leisure usage ($0.37–1.16 billion per year) [102].

According to Ahtiainen et al. [103, 104] the attempts to mitigate eutrophication levels for the Baltic Sea are related to both improvements in human welfare and economic benefits to coastal-resided citizens, as well as to the cost of degradation in preserving good environmental conditions for the nearby nature. In particular, the monetary advantages that occurred through eutrophication control were evaluated in a Baltic-wide stated preference contingent valuation study in 2014 [105]. The key aspects of describing eutrophication changes and achieving good water quality status in the Baltic Sea were based on citizens' stated willingness to pay. The relevant study aimed at reaching the target status while understanding a variety of eutrophication effects, such as that of water clarity, cyanobacterial blooms, underwater meadows, fish species composition, and oxygen deficiency at the sea bottom [105].

The aforementioned study covered the conditional changes reported among all nine coastal countries of the entire Baltic Sea. The target state was approached toward achieving a good environmental status of the sea, while all sub-basins except the Northern Baltic Proper, indeed, attained good status. The time frame in the study is interestingly longer than similar policy sets, reaching the year 2050. Nevertheless, taken into consideration that people are generally appreciating as more valuable those services and goods obtain sooner than later, a good status before 2050 might induce even greater benefits.

On the other hand, in an unsuccessful missing control of eutrophication in the Baltic Sea, there should be financial losses, as they are represented in Fig. 1.8. Consequently, total annual damages are expected at 3.8–4.4 billion € for the Baltic Sea region. Similarly, citizens' welfare can be increased by the relevant amounts of money each year if good eutrophication status can be achieved [105].

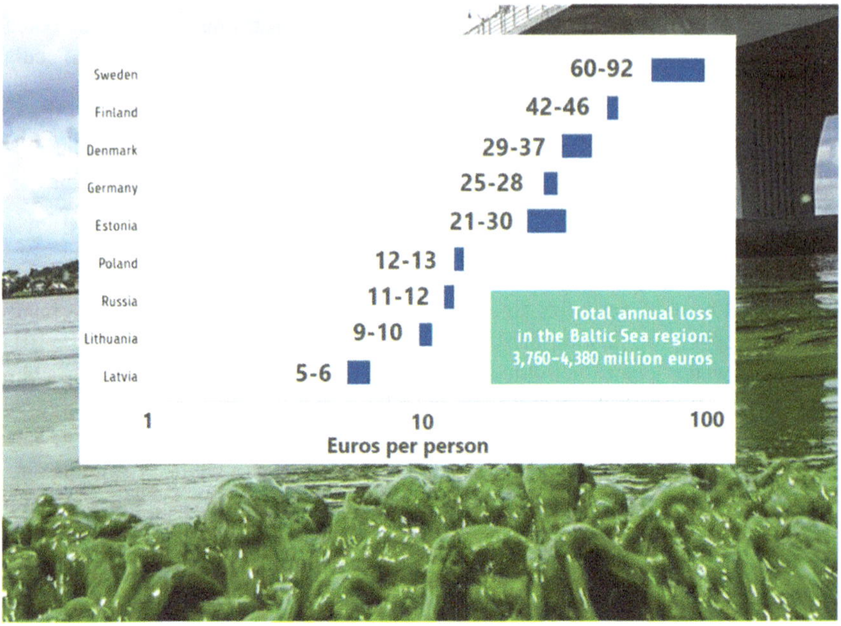

Fig. 1.8 Annual losses from eutrophication—euros per person and total—in the Baltic Sea. *Source* Modified from [105]

1.8 Conclusions

Based on the aforementioned introductory chapter of the book, the description of the eutrophication phenomenon was developed, while the generating causes of the phenomenon and its effects on the natural environment have been investigated. Despite the advancements of today scientific knowledge about elemental cycles and functionality in nature, cyanobacteria mechanisms functioning in natural water sources, as well as the theoretical background of eutrophication causes and results, many issues remain unanswered regarding the incidence, the environmental triggers for toxicity, the ability of timely prediction of phenomena duration, as well as the toxicity remediation measures. In the better understanding of environmental factors and toward building credible warning systems and prognostic tools that play an important role in decision making, the following two-fold approach has been recommended:

- At a micro-scale approach it is included the detection of causal factors of environmental fate and transportation, the deepening into the ecological processes concurred, as well as the development of laboratory methods to detect the underlying eutrophication simulations that should quantify the aforementioned theoretical approaches in practice.
- At a macro-scale approach, two types of interventions can be identified: (a) precautionary-proactive and (b) post-treatment. In particular, the precautionary

and the proactive measures and policies are mainly based on the environmental criteria of eutrophication causes, including the elemental cycles in nature, the weather and climatic conditions, the seasonal variations, and the human-induced interventions on water sources. The post-treatment measures and policies can empirically acknowledge the pragmatic sources of eutrophication, being based on the already accumulated experience of industrial and greywater wastewater treatment. These measures' and policies' scaling-up could be equally and satisfactorily applied: from the micro-scale toward the lake-, macro-scale, context accordingly.

Conclusively, the coordination and the integration of the aforementioned two scales remain challenging issues, while the economic dimension of healing the eutrophication-suffered water sources at a large-scale reality cannot be undermined.

References

1. Chislock M, Doster E (2013) Eutrophication: causes, consequences, and controls in aquatic ecosystems. Nat Educ Knowl
2. Khan MN, Mohammad F (2014) Eutrophication: challenges and solutions. In: Eutrophication: causes, consequences and control. ISBN 9789400778146
3. Lemley DA, Adams JB (2018) Eutrophication. In: Encyclopedia of ecology. ISBN 9780444641304
4. Chamie J (2020) World population: 2020 overview I YaleGlobal Online
5. OECD (2019) OECD work on environment
6. Daby D, Turner J, Jago C (2002) Microbial and nutrient pollution of coastal bathing waters in Mauritius. Environ Int
7. Ansari AA, Gill SS, Lanza GR, Rast W (2011) Eutrophication: causes, consequences and control. ISBN 9789048196241
8. Zamparas M, Zacharias I (2014) Restoration of eutrophic freshwater by managing internal nutrient loads. A review. Sci Total Environ
9. Smith VH, Schindler DW (2009) Eutrophication science: where do we go from here? Trends Ecol Evol 24:201–207
10. Smith VH (2003) Eutrophication of freshwater and coastal marine ecosystems: a global problem. Environ Sci Pollut Res
11. Correll DL (1998) The role of phosphorus in the eutrophication of receiving waters: a review. J Environ Qual
12. Anderson DM, Glibert PM, Burkholder JM (2002) Harmful algal blooms and eutrophication: nutrient sources, composition, and consequences. Estuaries
13. Carpenter SR (2008) Phosphorus control is critical to mitigating eutrophication. Proc Natl Acad Sci USA
14. Schindler DW (1977) Evolution of phosphorus limitation in lakes. Science (80-)
15. Schindler DW, Hecky RE, Findlay DL, Stainton MP, Parker BR, Paterson MJ, Beaty KG, Lyng M, Kasian SEM (2008) Eutrophication of lakes cannot be controlled by reducing nitrogen input: results of a 37-year whole-ecosystem experiment. Proc Natl Acad Sci U S A
16. National Research Council (2000) Sources of nutrient inputs to estuaries and coastal waters. In: Clean coastal waters: understanding and reducing the effects of nutrient pollution. ISBN 978-0-309-06948-9
17. Guan Q, Feng L, Hou X, Schurgers G, Zheng Y, Tang J (2020) Eutrophication changes in fifty large lakes on the Yangtze Plain of China derived from MERIS and OLCI observations. Remote Sens Environ

18. Peppa M, Vasilakos C, Kavroudakis D (2020) Eutrophication monitoring for Lake Pamvotis, Greece, using sentinel-2 data. ISPRS Int J Geo Inf
19. Smith VH, Joye SB, Howarth RW (2006) Eutrophication of freshwater and marine ecosystems. In: Proceedings of the limnology and oceanography
20. Dodds WK, Smith VH (2016) Nitrogen, phosphorus, and eutrophication in streams. Inland Waters
21. Smith VH (2009) Eutrophication. In: Encyclopedia of inland waters. ISBN 9780123706263
22. Das A, Mondal M, Das B, Ghosh AR (2012) Analysis of drainage morphometry and watershed prioritization in Bandu Watershed, Purulia, West Bengal through remote sensing and GIS technology—a case study Int J Geomat Geosci
23. Mondal S, Mukherjee AK, Senapati T, Pal S, Haque S, Ghosh AR (2015) Stratification and water quality of an abandoned opencast coal pit lake at Raniganj Coalfield Area, West Bengal, India. Lakes Reserv Res Manag
24. Ghosh TK, Mondal D (2012) Eutrophication: causative factors and remedial measures. J Today's Biol Sci Res Rev 1:153–178
25. Zacharias I, Zamparas M (2010) Mediterranean temporary ponds. A disappearing ecosystem. Biodivers Conserv
26. Brönmark C, Hansson LA (2017) The biology of lakes and ponds. ISBN 9780198713593
27. Katsev S (2016) Phosphorus effluxes from lake sediments
28. Biggs J, Williams P, Whitfield M, Nicolet P, Weatherby A (2005) 15 years of pond assessment in Britain: results and lessons learned from the work of Pond Conservation. In: Proceedings of the aquatic conservation: marine and freshwater ecosystems
29. Oertli B, Joye DA, Castella E, Juge R, Lehmann A, Lachavanne JB (2005) PLOCH: a standardized method for sampling and assessing the biodiversity in ponds. In: Proceedings of the aquatic conservation: marine and freshwater ecosystems
30. Brönmark C, Hansson LA (2002) Environmental issues in lakes and ponds: current state and perspectives. Environ Conserv
31. Hamerlík L, Svitok M, Novikmec M, Očadlík M, Bitušík P (2014) Local, among-site, and regional diversity patterns of benthic macroinvertebrates in high altitude waterbodies: do ponds differ from lakes? Hydrobiologia
32. Hamerlík L, Svitok M, Novikmec M, Veselská M, Bitušík P (2017) Weak altitudinal pattern of overall chironomid richness is a result of contrasting trends of subfamilies in high-altitude ponds. Hydrobiologia
33. Moosmann L, Gächter R, Müller B, Wüest A (2006) Is phosphorus retention in autochthonous lake sediments controlled by oxygen or phosphorus? In: Proceedings of the limnology and oceanography
34. Correll DL (1999) Phosphorus: a rate limiting nutrient in surface waters. Poult Sci
35. Wang S, Jin X, Pang Y, Zhao H, Zhou X, Wu F (2005) Phosphorus fractions and phosphate sorption characteristics in relation to the sediment compositions of shallow lakes in the middle and lower reaches of Yangtze River region, China. J Colloid Interface Sci
36. Spears BM, Carvalho L, Perkins R, Kirika A, Paterson DM (2007) Sediment phosphorus cycling in a large shallow lake: spatio-temporal variation in phosphorus pools and release. In: Proceedings of the hydrobiologia
37. Gianni A, Zamparas M, Papadas IT, Kehayias G, Deligiannakis Y, Zacharias I (2013) Monitoring and modeling of metal concentration distributions in anoxic basins: Aitoliko Lagoon, Greece. Aquat Geochem
38. Katsev S, Tsandev I, L'Heureux I, Rancourt DG (2006) Factors controlling long-term phosphorus efflux from lake sediments: exploratory reactive-transport modeling. Chem Geol
39. de Vicente I, Cruz-Pizarro L, Rueda FJ (2009) Sediment resuspension in two adjacent shallow coastal lakes: controlling factors and consequences on phosphate dynamics. Aquat Sci
40. Smolders AJP, Lamers LPM, Lucassen ECHET, Van Der Velde G, Roelofs JGM (2006) Internal eutrophication: how it works and what to do about it—a review. Chem Ecol
41. Burger DF, Hamilton DP, Pilditch CA (2008) Modelling the relative importance of internal and external nutrient loads on water column nutrient concentrations and phytoplankton biomass in a shallow polymictic lake. Ecol Model

42. Janssen ABG, de Jager VCL, Janse JH, Kong X, Liu S, Ye Q, Mooij WM (2017) Spatial identification of critical nutrient loads of large shallow lakes: implications for Lake Taihu (China). Water Res
43. Raheem A, Sikarwar VS, He J, Dastyar W, Dionysiou DD, Wang W, Zhao M (2018) Opportunities and challenges in sustainable treatment and resource reuse of sewage sludge: a review. Chem Eng J
44. Eurostat Sewage sludge production and disposal
45. FAO (2017) World fertilizer trends and outlook to 2020. Food and Agriculture Organization of the United Nations
46. Nikolai SJ, Dzialowski AR (2014) Effects of internal phosphorus loading on nutrient limitation in a eutrophic reservoir. Limnologica
47. Zamparas M, Kapsalis VC, Kanteraki AE, Vardoulakis E, Kyriakopoulos GL, Drosos M, Kalavrouziotis IK (2019) Novel composite materials as P-adsorption agents and their potential application as fertilizers. Glob Nest J
48. Me W, Hamilton DP, McBride CG, Abell JM, Hicks BJ (2018) Modelling hydrology and water quality in a mixed land use catchment and eutrophic lake: effects of nutrient load reductions and climate change. Environ Model Softw
49. Hickey CW, Gibbs MM (2009) Lake sediment phosphorus release management—decision support and risk assessment framework. N Z J Mar Freshw Res 43:819–856
50. Huser BJ, Futter M, Lee JT, Perniel M (2016) In-lake measures for phosphorus control: the most feasible and cost-effective solution for long-term management of water quality in urban lakes. Water Res
51. Lindim C, Becker A, Grüneberg B, Fischer H (2015) Modelling the effects of nutrient loads reduction and testing the N and P control paradigm in a German shallow lake. Ecol Eng
52. Djihouessi MB, Aina MP (2018) A review of hydrodynamics and water quality of Lake Nokoué: current state of knowledge and prospects for further research. Reg Stud Mar Sci
53. Zhu Y, McCowan A, Cook PLM (2017) Effects of changes in nutrient loading and composition on hypoxia dynamics and internal nutrient cycling of a stratified coastal lagoon. Biogeosciences
54. Isles PDF, Giles CD, Gearhart TA, Xu Y, Druschel GK, Schroth AW (2015) Dynamic internal drivers of a historically severe cyanobacteria bloom in Lake Champlain revealed through comprehensive monitoring. J Gt Lakes Res
55. Jeppesen E, Søndergaard M, Jensen HS, Ventäla AM (2009) Lake and reservoir management. In: Encyclopedia of inland waters. ISBN 9780123706263
56. Søndergaard M, Jeppesen E, Jensen JP, Amsinck SL (2005) Water framework directive: ecological classification of Danish lakes. J Appl Ecol
57. Kagalou I, Papastergiadou E, Leonardos I (2008) Long term changes in the eutrophication process in a shallow Mediterranean lake ecosystem of W. Greece: response after the reduction of external load. J Environ Manag 87:497–506
58. Kagalou I, Leonardos I (2009) Typology, classification and management issues of Greek lakes: implication of the Water Framework Directive (2000/60/EC). In: Proceedings of the environmental monitoring and assessment
59. Yang L, Lei K, Yan W, Li Y (2013) Internal loads of nutrients in Lake Chaohu of China: implications for lake eutrophication. Int J Environ Res
60. Gürevin C, Erturk A, Albay M (2017) Predicting the effects of sediment based internal nutrient loads on eutrophication in Küçükçekmece Lagoon for rehabilitation planning. Int J Sediment Res
61. Sánchez-Carrillo S, Álvarez-Cobelas M (2001) Nutrient dynamics and eutrophication patterns in a semi-arid wetland: the effects of fluctuating hydrology. Water Air Soil Pollut
62. Søndergaard M, Jensen JP, Jeppesen E (2003) Role of sediment and internal loading of phosphorus in shallow lakes. In: Proceedings of the hydrobiologia
63. Horppila J, Holmroos H, Niemistö J, Massa I, Nygrén N, Schönach P, Tapio P, Tammeorg O (2017) Variations of internal phosphorus loading and water quality in a hypertrophic lake during 40 years of different management efforts. Ecol Eng

64. Sánchez-Carrillo S, Angeler DG, Álvarez-Cobelas M, Sánchez-Andrés R (2011) Freshwater wetland eutrophication. In: Eutrophication: causes, consequences and control. ISBN 9789048196241
65. Wu Y, Wen Y, Zhou J, Wu Y (2014) Phosphorus release from lake sediments: effects of pH, temperature and dissolved oxygen. KSCE J Civ Eng
66. Rydin E (2000) Potentially mobile phosphorus in Lake Erken sediment. Water Res
67. Christophoridis C, Fytianos K (2006) Conditions affecting the release of phosphorus from surface lake sediments. J Environ Qual
68. Ogdahl ME, Steinman AD, Weinert ME (2014) Laboratory-determined phosphorus flux from lake sediments as a measure of internal phosphorus loading. J Vis Exp
69. Zamparas M, Drosos M, Deligiannakis Y, Zacharias I (2015) Eutrophication control using a novel bentonite humic-acid composite material Bephos™. J Environ Chem Eng 3:3030–3036
70. Bishop WM, Richardson RJ (2018) Influence of Phoslock® on legacy phosphorus, nutrient ratios, and algal assemblage composition in hypereutrophic water resources. Environ Sci Pollut Res
71. Amirbahman A, Lake BA, Norton SA (2013) Seasonal phosphorus dynamics in the surficial sediment of two shallow temperate lakes: a solid-phase and pore-water study. Hydrobiologia
72. Kyriakopoulos GL, Kapsalis VC, Aravossis KG, Zamparas M, Mitsikas A (2019) Evaluating circular economy under a multi-parametric approach: a technological review. Sustainability
73. Hupfer M, Hilt S (2008) Lake restoration. In: Encyclopedia of ecology, five-volume set. ISBN 9780080914565
74. Zamparas M, Deligiannakis Y, Zacharias I (2013) Phosphate adsorption from natural waters and evaluation of sediment capping using modified clays. Desalin Water Treat
75. Zamparas M, Drosos M, Kalavrouziotis IK (2019) Sorption of phosphate from innovative composite material focusing on physicochemical interactions. Desalin Water Treat 151:212–218
76. Kiani M, Tammeorg P, Niemistö J, Simojoki A, Tammeorg O (2020) Internal phosphorus loading in a small shallow lake: response after sediment removal. Sci Total Environ
77. Bennion H, Moss B, Jeppesen E, Carvalho L, Demeester L, Feuchtmayr H, Friberg N, Gessner MO, Hefting M, Lauridsen TL et al (2010) Interaction of climate change and eutrophication. In: Climate change impacts on freshwater ecosystems. ISBN 9781405179133
78. Hickey CW, Gibbs MM (2009) Lake sediment phosphorus release management—decision support and risk assessment framework. N Z J Mar Freshw Res
79. Phillips G, Bramwell A, Pitt J, Stansfield J, Perrow M (1999) Practical application of 25 years' research into the management of shallow lakes. In: Proceedings of the hydrobiologia
80. Heisler J, Glibert PM, Burkholder JM, Anderson DM, Cochlan W, Dennison WC, Dortch Q, Gobler CJ, Heil CA, Humphries E et al (2008) Eutrophication and harmful algal blooms: a scientific consensus. Harmful Algae
81. Grattan LM, Holobaugh S, Morris JG (2016) Harmful algal blooms and public health. Harmful Algae
82. Davidson K, Anderson DM, Mateus M, Reguera B, Silke J, Sourisseau M, Maguire J (2016) Forecasting the risk of harmful algal blooms. Harmful Algae
83. Mitchell I, Daunt F, Frederiksen M, Wade K (2020) Impacts of climate change on human health, HABs and bathing waters, relevant to the coastal and marine environment around the UK. MCCIP Sci Rev
84. Anderson DM (2019) Harmful algal blooms. In: Encyclopedia of ocean sciences. ISBN 9780128130810
85. O'Neil JM, Davis TW, Burford MA, Gobler CJ (2012) The rise of harmful cyanobacteria blooms: the potential roles of eutrophication and climate change. Harmful Algae
86. Huisman J, Codd GA, Paerl HW, Ibelings BW, Verspagen JMH, Visser PM (2018) Cyanobacterial blooms. Nat Rev Microbiol
87. Merel S, Walker D, Chicana R, Snyder S, Baurès E, Thomas O (2013) State of knowledge and concerns on cyanobacterial blooms and cyanotoxins. Environ Int

88. Nishikawa T, Hori Y, Nagai S, Miyahara K, Nakamura Y, Harada K, Tada K, Imai I (2014) Long-term (36-year) observations on the dynamics of the fish-killing raphidophyte Chattonella in Harima-Nada, eastern Seto Inland Sea, Japan. J Oceanogr
89. Tada K, Pithakpol S, Montani S (2004) Seasonal variation in the abundance of Noctiluca scintillans in the Seto Inland Sea, Japan. Plankton Biol Ecol
90. Imai I, Yamaguchi M, Hori Y (2006) Eutrophication and occurrences of harmful algal blooms in the Seto Inland Sea, Japan. Plankton Benthos Res
91. Imai I, Kimura S, Yamamoto T, Tomaru Y, Nagasaki K, Sakurada K, Murata K (2009) Possible prevention strategie for red tides of the fish-killer dinoflagellate Cochlodinium polykrikoides using microorganisms. Bull Plankton Soc Jpn
92. Okaichi T (1989) Red tide problems in the Seto Inland sea, Japan. In: Red tides: biology, environmental science and toxicology
93. Imai I, Yamaguchi M (2012) Life cycle, physiology, ecology and red tide occurrences of the fish-killing raphidophyte Chattonella. Harmful Algae
94. Dale B, Thorsen TA, Fjellså A (1999) Dinoflagellate cysts as indicators of cultural eutrophication in the Oslofjord, Norway. Estuar Coast Shelf Sci
95. Alves-de-Souza C, Pecqueur D, Le Floc'H E, Mas S, Roques C, Mostajir B, Vidussi F, Velo-Suárez L, Sourisseau M, Fouilland E et al (2015) Significance of plankton community structure and nutrient availability for the control of dinoflagellate blooms by parasites: a modeling approach. PLoS One
96. Pereira P, Botelho MJ, Cabrita MT, Vale C, Moita MT, Gonçalves C (2012) Winter-summer nutrient composition linkage to algae-produced toxins in shellfish at a eutrophic coastal lagoon (óbidos lagoon, Portugal). Estuar Coast Shelf Sci
97. Gianni A, Zacharias I (2011) Morphology and anoxia in enclosed seas: Aitoliko Lagoon—Greece. Geophys Res Abstr
98. Breitburg DL, Rose KA, Cowan JH (1999) Linking water quality to larval survival: predation mortality of fish larvae in an oxygen-stratified water column. Mar Ecol Prog Ser
99. U.S. EPA Office of Water (2015) Recommendations for public water systems to manage cyanotoxins in drinking water. ISBN 9788578110796
100. Graham JL, Dubrovsky NM, Eberts SM (2017) Cyanobacterial harmful algal blooms and U.S. geological survey science capabilities. USGS
101. Dodds WK, Bouska WW, Eitzmann JL, Pilger TJ, Pitts KL, Riley AJ, Schloesser JT, Thornbrugh DJ (2009) Policy analysis: eutrophication of U.S. freshwaters: damages. Environ Sci Technol 43:8
102. Dodds WK, Bouska WW, Eitzmann JL, Pilger TJ, Pitts KL, Riley AJ, Schloesser JT, Thornbrugh DJ (2009) Eutrophication of U.S. freshwaters: analysis of potential economic damages. Environ Sci Technol
103. Ahtiainen H, Vanhatalo J (2012) The value of reducing eutrophication in European marine areas—a Bayesian meta-analysis. Ecol Econ
104. Ahtiainen H, Artell J, Czajkowski M, Hasler B, Hasselström L, Huhtala A, Meyerhoff J, Smart JCR, Söderqvist T, Alemu MH et al (2014) Benefits of meeting nutrient reduction targets for the Baltic Sea—a contingent valuation study in the nine coastal states. J Environ Econ Policy
105. Ahtiainen H, Artell J, Elmgren R, Hasselström L, Håkansson C (2014) Baltic sea nutrient reductions—what should we aim for? J Environ Manag

Chapter 2
Eutrophication and Public Health

Apostolos Vantarakis

Abstract While there is no evidence of directly linking toxic algal blooms with eutrophication, it is highly potential biotoxins to endanger public health, having toxic algal blooms and red tides in eutrophic waters. Consequently, public health risk is associated with eutrophication through consumption of sea food contaminated with red tide toxins. Toxins contained in cyanophyte cells have also acute effects on vertebrates, from minor skin irritations, liver injury, impaired system to death of mammals, life, and pets. Toxic organisms in the water bodies result in bans on finfish and shellfish collection and commercialization, linked to economic losses at tourism and fish farming. Main categories of shellfish poisoning caused by toxic algae were denoted, while the impacts of eutrophic waters on waterborne pathogens in the marine environment were approached in terms of inadequately treated sewage, although several human bacterial pathogens are indigenous to estuarine and sea waters. Eubacteria and eutrophication were determined by: (a) many eubacteria possess gas vesicles that permit them to migrate vertically at intervals the water column, offering an excellent advantage over different flora in stratified waters, and, (b) many international recreational and drink sources suffer from eutrophication and eubacteria due to increased stream regulation, thus, conditioning nutrients and growth of eubacteria.

Keywords Biotoxins · Cyanophyte cells · Eubacteria · Eutrophication · Finfish · Live injury · Pathogens · Shellfish · Skin irritation · Toxins

i. Terminology

Anoxia	Lack of dissolved oxygen in water
Biomanipulation	The alteration of a food web to restore ecosystem health
Bloom	is an increase in abundance of a particular organism relative to its normal background level

A. Vantarakis (✉)
Department of Public Health, Medical School, University of Patras, Patras, Greece
e-mail: avanta@upatras.gr

M. G. Zamparas and G. L. Kyriakopoulos (eds.), *Chemical Lake Restoration*,
https://doi.org/10.1007/978-3-030-76380-0_2

	(which may be "low" or "high" depending on the species)
Chlorophyll a	The green pigment in plants that promotes photosynthesis
Eutrophication	elevated primary production
External loading	Nutrients from outside the lake, for instance, nutrients in wastewater discharge to the lake, or nutrients from agricultural drainage water
HAB	Harmful algal bloom; abundant phytoplankton
Hypoxia	Reduced dissolved oxygen concentration in water that stresses an organism
Internal loading	Release of nutrients, such as phosphorus and nitrogen, from sediments during low oxygen concentration conditions
Mechanical-biological treatment	Treatment of wastewater by settling of suspended matter together with the use of microorganisms to decompose organic matter. Only minor amounts of nutrients (nitrogen and phosphorus) are removed by this treatment
Nonpoint source (diffuse) pollution	Pollution mainly from agriculture or dumping grounds. It is difficult to collect for treatment
Photosynthesis	Conversion of inorganic carbon (carbon dioxide) to organic carbon (glucose) by a primary producer
Phytoplankton	Free-floating microscopic plants
Point source pollution	Polluted water from a defined point. It can be collected as industrial or municipal wastewater and treated by what is often called end-of-pipe technology (environmental technology)
Sediment	The bottom and often muddy layer of a lake
Thermocline	The level dividing a lake into two layers, an upper warmer one (epilimnion) and a lower colder one (hypolimnion). The temperature usually drops several degrees centigrade over just a few meters at this level
Zooplankton	Microscopic to very small free-floating animals

ii. **Key messages**

- Eutrophication—the over-enrichment of waters by nutrients—threatens and degrades many coastal and freshwater ecosystems around the world.
- The two most acute symptoms of eutrophication are hypoxia (or oxygen depletion) and harmful algal blooms, which among other things can destroy aquatic life in affected areas.

- Of all areas around the world identified as experiencing some form of eutrophication, several could be characterized as systems in recovery if measures will be undertaken.
- Mapping and research into the extent of eutrophication and its effects on human health and ecosystem services has been improved, but there is still insufficient information in many regions of the world to establish the actual extent of eutrophication or identify the sources of nutrients.
- Data from epidemiologic studies designed to carefully evaluate the associations among environmental toxin concentrations, exposure routes, and related health symptoms are needed to develop environmental contaminant exposure guidelines for both people and animals.
- To develop effective policies and strategies to mitigate eutrophication, more information is required on the extent of eutrophication, the sources of nutrients, and the impact of eutrophication on ecosystems.

2.1 Introduction

Eutrophication is the main cause for impairment of fresh and coastal marine ecosystems within the world. It occurs when several nutrients—like phosphates and nitrates—lead to the aquatic environment due to human activities and is mostly related to the anthropogenetic pollution of water with excessive nutrients. Eutrophication of natural water bodies is well documented because of the growth of the productivity of the water body, as well as water lots, bottom sediments, organisms inhabiting them and therefore the relations between them. It may be a fact, that is achieved by enrichment with nutrients, that stimulates the expansion and flowering of alga, and this, in turn, results in a deterioration within the quality and condition of natural waters. In recent decades, this has become a world environmental drawback [3, 5, 7, 16, 20, 30, 53]. The impact of eutrophication can be the rapid increase in biomass, which may have negative and a few positive effects. Nevertheless, mainly, the eutrophication of water bodies establishes a negative impact on human health, causing public health issues. Nutrient sources could be either *point* sources or *diffuse* sources (Fig. 2.1).

These nutrients originate from the agriculture, the factories and therefore the waste matter treatment plants (Fig. 2.2). These phosphates and nitrates can be found, in turn, within the ocean via the rivers.

Positive effects relate to the utilization of the water body as a supply of chemicals for economic profit, like for fish farms or rice paddies. The negative effects include increased turbidity, dominance of species and thus loss of biodiversity, disturbing odor, style, and color of water bodies inappropriate for tourism, issues for drink treatment and risk of poisons inflicting human and environmental health issues.

Many eubacteria possess gas vesicles which permit them to migrate vertically at intervals of the water column to succeed in optimum growth conditions. This facility will give them an excellent advantage over different flora, significant in stratified

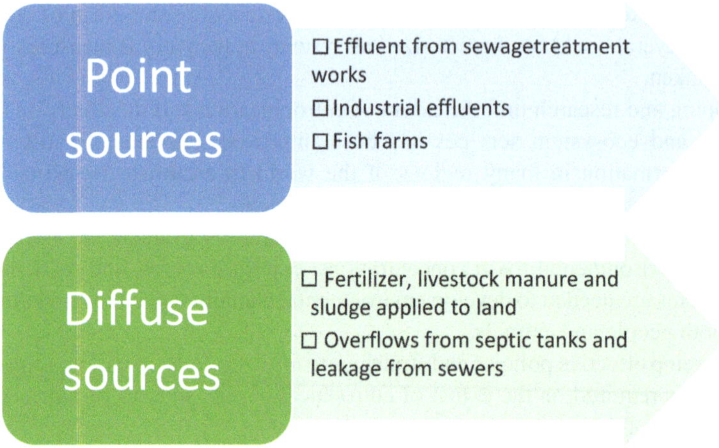

Fig. 2.1 Types of nutrient sources

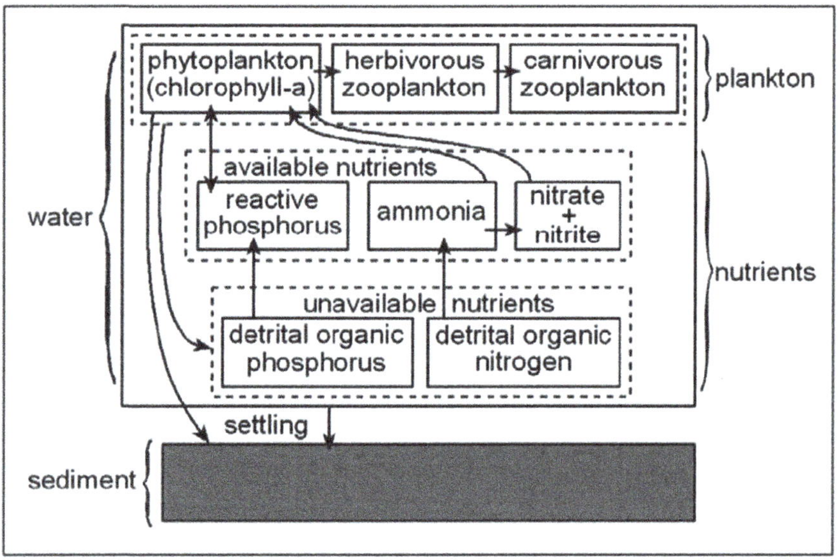

Fig. 2.2 Available via CC BY-NC-ND 4.0

waters. Even in well-mixed, shallow waters, eubacteria could have an advantage because they are capable of growth in variable levels of sunshine and with low levels of phosphorus. Some species, additionally, proliferate in waters having an affinity for phosphorus, nitrogen, and greenhouse gas. Many international recreational and drink sources suffer from eutrophication and outbreaks of eubacteria, principally due to increased stream regulation. This has provided ideal conditions for concentrations of nutrients and growth of eubacteria. The toxins contained in cyanophyte cells have acute effects on vertebrates, from minor skin irritations, liver injury, impaired system to death of mammals, life, and pets.

Eutrophication is characterized by excessive plant and protist growth because of the increased availability of one or a few limiting growth factors required for chemical process [66], like daylight, greenhouse gas, and nutrient fertilizers. Eutrophication happens naturally over centuries as lakes age and crammed in with sediments [13, 15, 17, 18]. However, human activities have accelerated the speed and extent of eutrophication through each point-source discharges and non-point loadings of limiting nutrients, like phosphorus, into aquatic ecosystems (i.e., cultural eutrophication), with dramatic consequences for drink sources, fisheries, and recreational water bodies [19, 22–24]. As an example, scientists and pool managers usually deliberately eutrophic water bodies by adding fertilizers to boost primary productivity and increase the density and biomass of recreationally and economically necessary fishes via bottom-up effects on higher biological process levels [12]. However, throughout the 1960 and 1970, scientists connected protist blooms to nutrient enrichment from anthropogenetic activities like agriculture, industry, and disposal [65] (Fig. 2.3).

The acknowledged consequences of eutrophication are to embrace blooms of cyanobacteria to drink water sources, degradation of recreational opportunities, and drive. A rough estimate is that the calculable price of harm mediate by eutrophication within the U.S. alone is roughly $2.2 billion annually.

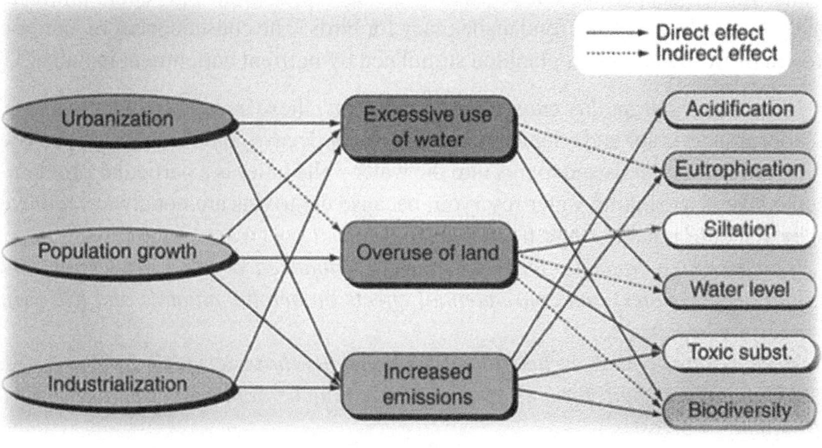

Fig. 2.3 The basic factors which cause environmental problems in lakes and reservoirs. *Source* modified from Farley [32] with permission from Springer Nature

2.2 Consequences

Eutrophication is one of the most prevalent challenges faced by freshwater systems globally [48, 67, 72] and has a long historical background in terms of water management [66]. Eutrophication leads to major disruptions of aquatic ecosystems, affecting related goods and services [27], ecosystem and human health [46], as well as economic activities [63]. Based on this context, harmful algal blooms represent one of the major ecological health risks associated with eutrophication [59]. This occurs for cyanobacteria [60] that produce harmful materials, such as toxins [26, 54] and flavor substances [43, 73], which are potential hazards for both human and wildlife health. Nutrient enrichment can interact with site-specific conditions, such as the presence of other contaminants, including infectious disease agents, thereby threatening the ecological stability of the system. In addition, the accelerated eutrophication of aquatic ecosystems has also been associated with the emergence of new diseases through both direct and indirect pathways [41], emphasizing the importance of acknowledging and understanding the mechanisms linking eutrophication and host–pathogen interactions. To ensure the sustainability of freshwater systems, a better understanding, as well as assessment and control of eutrophication, is essential.

In recent decades, eutrophication has become a world environmental drawback [3, 16, 20, 30, 53]. Often, the issue of waters' eutrophication is considered solely from the purpose of degradation of water quality in an exceeding water body. Due to high range of algae, there's less lightweight within the water and therefore the plants situated deeper, die out because of the decomposition. This could ultimately cause the death of fish and food inadequacy for birds. The consequences of increased growth of freshwater phytoplankton stimulated by nutrient enrichment include [32]:

- *Effects on water quality caused by algal blooms:* these include changes to a lake's appearance, color and odor as a bloom grows or decays, and the release of colored organic compounds and toxins into the water—the latter is a particular problem if the lake is a drinking water reservoir, because the toxins are not always removed by standard potable water treatments.
- *Increased hypolimnetic deoxygenation (as enhanced organic production sinks into deeper water) and consequential effects on benthic animals and fish living near the lake bed.*
- *More general effects on lake food webs, such as those resulting from increased dominance of chain-forming cyanobacteria which zooplankters find noxious or difficult to ingest.*

Sometimes there has been explicit or unintentional confusion of the effects of plant nutrient enrichment and the effects of BOD addition (and the oxygen-consuming decay of organic material releasing nutrients which secondarily stimulate more organic production). The first can lead to the production of extra organic matter which consumes oxygen when it decays. The second involves the oxygen-consuming decay of organic material, which can release mineral nutrients that secondarily stimulate more organic production. It is preferable to use the term "eutrophic" to effects stimulated by plant nutrients. The distinction between *allochthonous* (external) and *autochthonous* (local) production is useful. BOD is the result of allochthonous production; nutrient enrichment stimulates autochthonous production.

The most conspicuous effect of eutrophication is the creation of dense blooms of noxious, foul-smelling phytoplankton that reduce water clarity and harm water quality. Algal blooms limit light penetration, reducing growth and causing die-offs of plants in littoral zones while also lowering the success of predators that need light to pursue and catch prey [44]. Furthermore, high rates of photosynthesis associated with eutrophication can deplete dissolved inorganic carbon and raise pH to extreme levels during the day. Elevated pH can in turn "blind" organisms that rely on the perception of dissolved chemical cues for their survival by impairing their chemosensory abilities [76]. When these dense algal blooms eventually die, microbial decomposition severely depletes dissolved oxygen, creating a hypoxic or anoxic "dead zone" lacking sufficient oxygen to support most organisms. Dead zones are found in many freshwater lakes including the Laurentian Great Lakes e.g., central basin of Lake Erie [2] during the summer. Furthermore, such hypoxic events are particularly common in marine coastal environments surrounding large, nutrient-rich rivers (e.g., Mississippi River and the Gulf of Mexico) and have been shown to affect more than 245,000 km^2 in over 400 near-shore systems [25]. Hypoxia and anoxia because of eutrophication continue to threaten lucrative commercial and recreational fisheries worldwide.

Eutrophication is one of the primary challenges to sustainable water management and needs urgent attention due to cumulative anthropogenic activities. It is increasingly recognized that eutrophication has multidimensional consequences linked to water quality, ecosystem, and human health, as well as economic activities. Furthermore, the enrichment of nutrients interacts with many site-specific conditions, specifically, the ecological stability of the system, land-use types, climate change, and the presence of other contaminants, including infectious disease agents. It is essential that future studies acknowledge and consider the complexity of these multiple interactions.

2.3 Why Eutrophication Is a Problem?

Phytoplankton or microalgae are photosynthesizing microscopic organisms. They contain pigment and need daylight to grow. Most plants float in the higher part of the ocean wherever daylight penetrates the water in an exceedingly balanced trophic system, and they constitute food for various organisms like whales, shrimp, snails, and jellyfish. Among these are the diatoms, eubacteria, dinoflagellates, and coccolithophores. Plant species need inorganic nutrients like nitrates, phosphates, and sulfur that they convert into proteins, fats, and carbohydrates. Blooms within the ocean might cowl an oversized space and are visible in satellite pictures. Harmful protist blooms (HAB) are bloom events involving harmful plants. These cause damage through the production of poisons or by their accumulated biomass, which may result in occurring organisms and alter food cycle dynamics.

Phytoplankton are the food supply for various alternative organisms, particularly the animals which are heterotrophic organism. They're primarily transported by close water currents however several will swim. Through their consumption and process of plant and alternative food sources, they play a role in aquatic food webs as a resource for higher biological process levels as well as fish. They are divided in 2 vital groups: crustacean (copepods and krill) and gelatin-like animals.

The growth of plants causes increased turbidity or minimized penetration of sunshine into the lower depths of the water column. In lakes and rivers, this may inhibit the growth of submerged aquatic plants and affect species that are eating with them such as fish, or shellfish. Protist blooms might cause robust fluctuations in oxygen levels. As the protist population is growing at a quick rate, it's going to block daylight from reaching alternative organisms and cause a decrease in oxygen levels. As protists die, they're rotten by microorganism that consumes oxygen and the water will become briefly hypoxic. The direct effects of this is, particularly, to cause the death of fish due to lack of dissolved oxygen. These hypoxic conditions promote the expansion of eubacteria because of increased phosphorus unleash. Also, several eubacteria species turn out toxins that are fatal to birds and animals.

Many observations and experiments provide evidence that the occurrence of so-called "green tides", consisting of marine seaweeds or macroalgae, is increased in eutrophicated coastal waters. These blooms are recognizable by large blades of algae that may wash up onto the shoreline. Green tides can asphyxiate aquacultures or disrupt traditional artisanal fisheries. Macroalgae can also be valorized as a raw material; harvesting could therefore be an effective solution [25–31, 34, 35].

The rise in eutrophic and hypoxic events has been primarily attributed to the rapid increase in intensive agricultural practices, industrial activities, and population growth, which together have increased nitrogen and phosphorus flows in the environment. Human activities have resulted in almost doubling nitrogen and tripling of phosphorus flows to the environment when compared to natural values [65]. By comparison, human activities have increased atmospheric concentrations of carbon dioxide, the gas primarily responsible for global warming, by approximately only 32% since the onset of the industrial age. Before nutrients—nitrogen in particular—are delivered to coastal ecosystems, they pass through a variety of terrestrial and freshwater ecosystems, causing other environmental problems such as freshwater quality impairments, acid rain, the formation of greenhouse gases, shifts in community food webs, and a loss of biodiversity. Once nutrients reach coastal systems, they can trigger several responses within the ecosystem. The initial impacts of nutrient increases are the excessive growth of phytoplankton, microalgae (e.g., epiphytes and microphytes), and macroalgae (i.e., seaweed). These can lead to other impacts described in Fig. 2.4.

The scientific community is increasing its knowledge of how eutrophication affects coastal ecosystems, yet the long-term implications of increased nutrient fluxes in our coastal waters are currently not entirely known or understood. Eutrophication diminishes the ability of coastal ecosystems to provide valuable ecosystem services

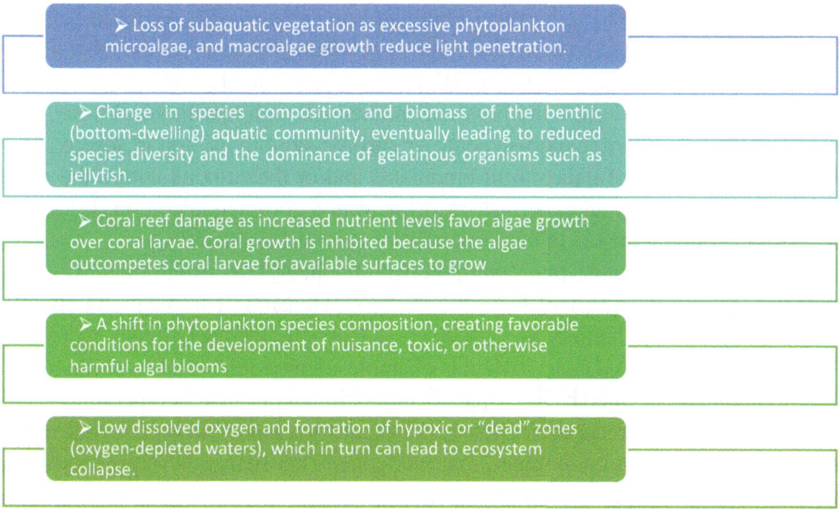

Fig. 2.4 Secondary impacts of eutrophication

such as tourism, recreation, the provision of fish and shellfish for local communities, sportfishing, and commercial fisheries. In addition, eutrophication can lead to reductions in local and regional biodiversity [34–38].

Today nearly half of the world's population lives within 60 km of the coast, with many communities relying directly on coastal ecosystems for their livelihoods. This means that a significant portion of the world's population is vulnerable to the effects of eutrophication in their local coastal ecosystems. Two of the most acute and commonly recognized symptoms of eutrophication are mainly harmful algal blooms and hypoxia.

A "Harmful algae Bloom" (HAB) could be a separate event related to a proliferation of microalgae, blue-green algae or macroalgae that are perceived by humans as harmful to their health or system services. Bloom is outlined as a rise in the abundance of a selected organism relative to its traditional noise level (which is also "low" or "high" looking on the species). As an example, excessive protist biomass accumulation will cause problems to aquatic organisms due to reduced daylight and to chemical element decrease within the water after bloom degradation. High biomass blooms of microalgae that synthesize ichthyotoxins or physically harm fish gills will cause large fish kills (in cultivation or natural fisheries). Alternatively, microalgae turn out potent unhealthful compounds that cause hurt to humans by direct contact, inhalation, or consumption of contaminated food. HAB can cause fish kills, human illness through shellfish poisoning, and death of marine mammals and shorebirds. They are often referred to as "red tides" or "brown tides" because of the appearance of the water when these blooms occur. One red tide event, which occurred near Hong Kong in 1998, wiped out 90% of the entire stock of Hong Kong's fish farms and resulted in an estimated economic loss of $40 million. Foodborne poisonings (mostly Paralytic

Shellfish Poisoning (PSP), diarrheal Shellfish Poisoning (DSP), Amnesic Shellfish Poisoning (ASP), and Ciguatera Fish Poisoning (CFP) [8] represent the most important threat to human health. The HAB toxins (thermostable and with chemicals terribly diverse) transferred and bio-concentrated through marine food webs up to humans UN agency will expertise acute or chronic symptoms, and in severe cases, death. The presence of unhealthful organisms and/or their toxins within the water end in bans on finfish and shellfish assortment and commercialization, leading to necessary economic losses in numerous sectors as well as business.

Global climate change may additionally influence HAB events, intensity, and impacts. Ocean warming will favor the biology extension of tropical and climatic zone organisms, like Gambierdiscus species, to a lot of temperate latitudes, increasing the threat of CFP. HAB is a particular example of phytoplankton (and phytobenthos) dynamics controlled by a complex combination of physical (temperature, wind, waves, atmosphere-ocean interactions), chemical (inorganic and organic nutrients), and biological (predation, competition, allelochemistry, parasitism) processes that operate at different spatio-temporal scales [8].

Harmful Algal Blooms occur in all aquatic environments and latitudes and because they are essentially natural phenomena, their occurrence cannot be completely avoided or eliminated. However, certain anthropogenic effects (e.g., eutrophication, alteration of the water circulation patterns by harbor construction, habitat destruction, spread of harmful organisms through ballast waters, or transport of cultured organisms) especially in the coastal zone, can benefit some HAB [47]. Technically, in the 21st Century, it is possible to minimize those activities, and to address the sustainable use of natural resources, which in turn will decrease in part the probabilities of HAB occurrence. The HAB toxins (thermostable and chemically very diverse) are transferred and bio-concentrated through marine food webs up to humans who can experience acute or chronic symptoms, and in severe cases, death. The presence of toxic organisms and/or their toxins in the water result in bans on finfish and shellfish collection and commercialization, resulting in important economic losses in different sectors including tourism [1]. Nevertheless, future trends are unsure [79], whereas the identification of latest toxins will complicate the risks to human health.

Hypoxia, considered to be the most severe symptom of eutrophication, has escalated dramatically over the past 50 years, increasing from about 10 documented cases in 1960 to at least 169 in 2007. Hypoxia occurs when algae and other organisms die, sink to the bottom, and are decomposed by bacteria, using the available dissolved oxygen [41, 46, 50, 53, 56]. Salinity and temperature differences between surface and subsurface waters lead to stratification, limiting oxygen replenishment from surface waters and creating conditions that can lead to the formation of a hypoxic or "dead" zone (Fig. 2.5).

Two of the most well-known hypoxic areas are the Gulf of Mexico and the Black Sea. The Gulf of Mexico has a seasonal hypoxic zone that forms every year in late summer. Its size varies; in 2000, it was less than 5,000 km^2, while in 2002 it was approximately 22,000 km^2 (or the size of the U.S. state of Massachusetts). While the economic consequences of the Gulf of Mexico dead zone are still unclear, concern over its increasing size led to the formation of the Mississippi River/Gulf of Mexico

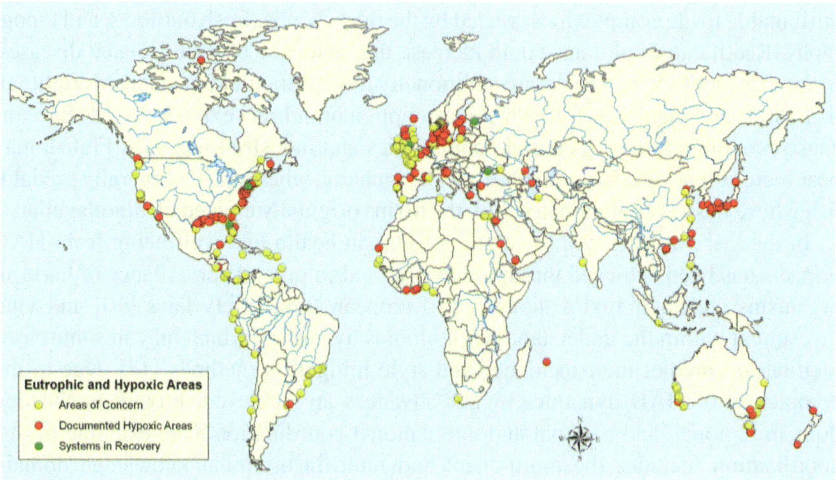

Fig. 2.5 Eutrophication and Hypoxia in coastal areas: a global assessment of the state of knowledge [68]

Watershed Nutrient Task Force in 1997 to develop a strategy to reduce the five-year running average areal extent of the Gulf of Mexico hypoxic zone to less than 5,000 km^2. The Black Sea, which was once the largest dead zone in the world, had 26 commercially viable fish species in the 1960s but eutrophication and hypoxia in Coastal Areas caused only five species by the 1980s. The growth of the Black Sea hypoxic zone was attributed to the intensified action of agriculture in the former Soviet Union. It has been "in recovery" since the economic collapse of Eastern Europe in the 1990s, which resulted in significant reductions in fertilizer use.

2.4 Human Health Impacts

There are risks from consumption of low doses of cyanotoxins, from water or bioaccumulation in crops irrigated with contaminated water. Marine blue-green algae found to harbor *Vibrio* to human or marine life known in European coastal waters which cause paralytic, toxin, and diarrhetic shellfish poisoning. Many protist species producing toxins harmful reaching dangerous levels for human and animal health. Harmful protist bloom species have the capability to produce community to combat these issues. However, more knowledge is needed to develop correct guidelines aimed toward property development, supported by tips to assist government agencies and thus the necessary warnings for recreational waters. The health risk could be reduced by physical removal of cells and poison from lakes with processes that ought to still take many years till the new equilibrium between sediment and water attributable to the complexities created by the tides, winds, fresh outflows, and topography. Rectification of dams could increase the incidence of water-related diseases. Marine cyanophyte outbreaks are additionally more difficult to manage. Mortality of fish, birds, and mammals following consumption or indirect exposure to HAB toxins cause economic losses to coastal communities and industrial fisheries. Finfish may be a vector for toxins, as within the case of ciguatera, wherever it's generally predator fish whose flesh is contaminated with the toxins originally created by dinoflagellates.

In the last 50 years, the protection of human health and well-being from HAB impacts has been achieved through intensive and improved surveillance of harmful organisms and their toxins along with European food safety laws [40] and vital investment within the understanding of blooms dynamics, which may in some cases facilitate to predict their incidence and style mitigation methods [14]. Due to the complexity of HAB dynamics, major advances are achieved through knowledge domain regional, and national and international coordination. One example of this coordination includes the multi-agent and multi-institutional knowledge domain collaboration over a few years to research the repeated large blooms of the neurolysin and fish-killing HAB, Karenia brevis (Florida red tide), within the Gulf of Mexico and American state. Progresses involved poison detection in water and air, and human health bar protocols and medicinal approaches.

An example of a project is the RAMOGE that coordinates the strategies of Monaco, Italy, France, and Spain for monitoring Ostreopsis blooms in some beaches where the proliferation of this benthic dinoflagellate is associated with mild respiratory distress in humans. RAMOGE also facilitates communication between scientists, stakeholders, and public. RAMOGE additionally facilitates communication between scientists, stakeholders, and public. There are risks of chronic illness from consumption of low doses of cyanotoxins, like from water or from bioaccumulation in crops irrigated with contaminated water. Marine cyanobacteria are found to harbor Vibrio to human or marine life are identified in European coastal waters. Several algal species able of manufacturing toxins harmful and more generally in seafood reaching dangerous levels for human and animal health. Examples include toxins dangerous to humans.

Algae toxins are observed in marine ecosystems where they'll accumulate in shellfish values and set suitable targets to regulate eutrophication. However, further research data is required to develop accurate guidelines aimed toward sustainable development, supported by guidelines to help government agencies and therefore the issuing of warnings for recreational waters. Several countries have developed legislation, and the health risk must be reduced by physical removal of cells and toxin by treatment. Although, lakes using these processes should still take several years until the new equilibrium between sediment and water because of the complexities created by the tides, winds, freshwater outflows, and topography. Remediation of dams may increase the incidence of water-related diseases. Marine cyanobacterial outbreaks are more difficult to manage [70, 72, 76].

Mortality of fish, birds, and mammals following consumption or indirect exposure to HAB toxins cause economic losses to coastal communities and commercial fisheries. Finfish can also be a vector for toxins, as in the case of ciguatera, where it is typically predator fish whose flesh is contaminated with the toxins originally produced by dinoflagellates. People are poisoned when consuming contaminated fish. Symptoms include gastrointestinal and neurological effects.

Nevertheless, future trends are uncertain, while the identification of new toxins can complicate the risks to human health (Berdalet et al. 2015). In the last 50 years, the protection of human health and wellbeing from HAB impacts has been achieved through intensive and improved monitoring of harmful organisms and their toxins coordinated with European food safety regulations and by significant investment in the understanding of blooms dynamics, which can in some cases help to predict their occurrence and design mitigation strategies [14]. Due to the complexity of HAB dynamics, major advances have been achieved through interdisciplinary regional, and national and international coordination [58, 63].

At global international scale, the SCOR and IOC UNESCO programs GEOHAB (2000–2013) and Global HAB (2016–2025) foster coordination on interdisciplinary, international, and interinstitutional research of HABs. Among the themes of Global HAB, that of "HABs and Human Health" aims to increase collaborations among scientists with medical, veterinary, public health, economic, and other social science expertise to help understand and minimize the risk of HAB impacts on human health and animal health, with a particular global change perspective. The overall approach followed by the research on HABs can inspire analogous strategies to address other challenges posed to oceans and human health.

Socio-cultural impacts (e.g., change on dietary patterns and traditional seafood—professional and recreational—collection activities) are particularly relevant in the case of CFP, endemic in tropical areas (Friedman et al. 2017), or due to exceptional blooms such as those of the domoic-acid (ASP) producer, Pseudonitzschia in 2015 in the Pacific coasts of America (Ritzman et al. 2018). Estimating the total cost of HABs is difficult and complex. A conservative total annual direct cost of HABs could account for about 100 billion US$, excluding costs for human health care [77, 82].

2.5 Drinking Water Risk

Several drinking water supplies throughout the world experience periodic massive surface blooms of cyanobacteria primarily caused by cultural eutrophication. These blooms contribute to a wide range of water-related problems, including fish kills and the formation of trihalomethane during water chlorination. Consumption of cyanobacterial blooms, or watersoluble neuro- and hepatoxins released when these blooms collapse can kill livestock and could pose serious health hazards to humans. For example, in Ethiopia, specifically Lake Chamo had experienced periodic fish kills, because of algal blooms, which was perhaps associated with cyanotoxin production. Willén et al. [80] have studied cyanotoxin production in seven rift valley lakes of Ethiopia and reported cyanobacterial blooms and occurrence of cyanotoxins (microcystins) in four of the studied lakes (Koka, Chamo, Langano and Ziway). The microcystins concentration in Lake Koka (45–51 μg/l) was the highest reported for African lakes and exceeds the levels that would cause health risks for humans, livestock, and wildlife [80]. The authors associated their year-round findings with the major cause for the death and poisonings of humans, domestic and wild animals around Lake Koka. The microcystin concentrations recorded in the studies are significantly higher than the provisional drinking water guideline value of 1 μg/l, recommended by WHO. Recently, an additional study was done on the concentration of microcystins in the liver and muscle tissue of commercially important fish species collected from Lake Koka [83]. The possible exposure routes are consumption of water, aquatic food, such as fish, crops, and vegetables contaminated with microcystins [21]. Oral ingestion of microcystin-contaminated water and/or foodstuffs has been associated with hepatocellular damage [21]. In some specific cases, local authorities must rely on eutrophic waters for producing drinking water.

The growth of algae in raw water reservoirs can cause problems for water treatment and supply, in relation to:

- Rapid blockage of filters, causing short filter runs and reduced throughput and/or increased volumes of wastewater from backwashing
- Increased dissolved organic carbon (DOC) in treated water from algal polysaccharide, which can result in customer complaints and increased chlorination by-product concentrations
- Penetration of algae through filters
- Taste and odor problems
- Production of toxins.

Treating raw water with high levels of organic matter is always technically difficult. It can lead to the creation of carcinogenic by-products (Trihalomethanes (THMs)), other chlorinated components or ozonides because of their reaction with disinfectants. If the water is eutrophic, then on top of the organic matter that would be present under normal circumstances, there will also be the organic matter produced by cyanobacteria (toxins and intracellular materials). An apparent association between bladder cancer and THMs has been demonstrated in an epidemiological study. However, as chlorinated water contains many by-products it is not possible from such epidemiological studies to conclude.

When faced with eutrophication of the water reservoir the best option, where possible, is to rely on an alternative resource for the water production. If it is not possible, then some changes can be made to the existing treatment chain, but there is no guarantee that the product will be totally safe. Information of the receiving population is necessary. Distribution of bottled water to the population at risk can be an option to consider. More detailed information on the changes that can be made to the treatment process and on the resistance of toxins to various disinfectants can be found in the technical annex. Figure 2.6 record the reasons for monitoring a water body for eutrophication. This is mostly relevant for water companies, which must monitor for eutrophication to treat eutrophic waters.

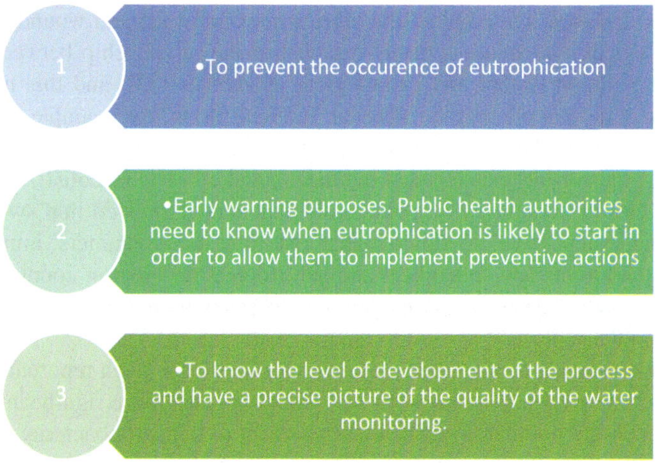

Fig. 2.6 Reasons for monitoring a water body for eutrophication

2.6 Public Health Risks

Biotoxins produced by toxic algae are of major public health concern worldwide. Although there is no evidence to directly link toxic algal blooms with eutrophication, there is little argument that the probability of having toxic algal blooms and red tides is much higher in eutrophic waters. Public health risk associated with eutrophication is mainly through consumption of sea food contaminated with red tide toxins, although direct contact with toxic algae may also cause dermatitis and conjunctivitis. Shellfish poisoning caused by toxic algae may be grouped into four major categories. The various types of shellfish poisoning have caused significant public health problems worldwide. For example, 24 people died and more than 1000 suffered from PSP in a Pyrodinium bloom in the Philippines [11]. 441 cases of PSP and 15 deaths were reported after consuming the hard clam Meritrix meritrix and the Sardinella spp. during three major PSP outbreaks in Indonesia [70]. In France, 3000 people were affected by DSP intoxication in 1983, after consumption of mussels exposed to a bloom of Dinophysis spp. [6]. In Hong Kong, levels of PSP toxins in shellfish collected from Tolo have tripled from 1984 to 1987 [50].

Another issue in eutrophic waters is waterborne pathogens in the marine environment which are mainly derived from untreated or inadequately treated sewage, although several human bacterial pathogens are indigenous to estuarine and sea waters (e.g., halophilic vibrios). Infection may be through consumption of contaminated seafood or direct contact through bathing or recreational water activities. Gastrointestinal tract infection, caused by water-borne pathogens remained one of the most widespread health problems in coastal cities with high population densities.

Contact with pathogenic bacteria (e.g., *Staphylococcus aureus*, *Pseudomonas aeruginosa*) and certain viruses in seawater may cause ear, eye, and skin infection or respiratory diseases, while halophilic vibrios may cause ear and wound infections. Epidemiological evidence has firmly established the relationship between bathing in sewage polluted waters and pathological incidences [33], and this means that potentially many people may be affected. In Hong Kong, for example, 42 beaches are visited by more than 20 million people each year. In 1997, the microbiological standard of these beaches exceeded the local standard (180 E. coli/100 ml), and 4 had to be closed. Results of an epidemiological survey showed that swimmers in Hong Kong waters have 1.5 times higher risk than non-swimmers, implying that many people are at risk. In tropical and sub-tropical developing countries, a very high percentage of sewage is discharged into the sea without treatment, and seafood harvested in these places constitutes a major health hazard.

Epidemics caused by the consumption of bivalves have been reported occasionally all over the world, although such problems are much more significant in developing and densely populated countries. High levels of indicator bacteria, pathogenic bacteria, and viruses are commonly found in bivalves, sediment, and water along the West African coast, in Asian seas, the Pacific and the Caribbean. Correspondingly, a high incidence and epidemic outbreaks of gastro-intestinal infections, diarrhea, and viral hepatitis caused by consumption of contaminated shellfish have been widely reported from these areas.

Hepatitis A Virus (HAV) is a major water-borne pathogen worldwide [51]. In 1988, a major outbreak of HAV occurred in Jiangsu and Jiangjang, China. About 400,000 people were infected due to the consumption of polluted shellfish (Acra spp.), and a dose-response relationship between consumption and HAV infection was clearly established [82]. In Hong Kong, 1458 cases of hepatitis were reported in the first three months of 1988. Of these, 647 cases were confirmed as Hepatitis A and some 80% of the patients had consumed local shellfish prior to infection [36]. The incidence of HAV ranged from 264 to 3636 incidences/year in the last ten years. In 1996, for example, 170 out of 264 (64%) of HAV infections were caused by the consumption of shellfish. Contamination of seafood is not only a problem restricted to developing countries, but also occurs in developed countries. For example, 57 outbreaks of viral gastroenteritis were reported in New South Wales, Australia in 1990, in which 1750 people were affected, 550 of them (31%) ate oysters [10]. A high level of enteric virus contamination in shellfish samples was also found from a major urban market in Europe, suggesting that the existing sanitary practices (culture, handling, and storage), as well as standards and regulation, need to be re-examined.

Regular monitoring of coastal water and shellfish is required to protect human health. However, water-borne pathogens normally occur in low numbers in water samples, and large errors may therefore be incurred in sampling and enumeration. Furthermore, culture techniques for specific pathogens are invariably species-specific, tedious, time-consuming and insensitive and molecular techniques are not expensive and require significant expertise. Because of these difficulties, pathogens are not normally enumerated in water quality and shellfish monitoring. Instead, fecal coliforms and E. coli which occur in high numbers in animal faces, are used to indicate fecal pollution, and hence the risk of potential exposure to pathogens derived from fecal origins.

Recent epidemiological evidence has shown that waterborne pathogens, as well as pathogenic incidence, do not correlate well with total coliform, fecal coliform, or E. coli counts. For example, the incidence of bathing-related GI diseases is much better correlated with Enterococcus than E. coli. and not related with total fecal coliform counts, or with fecal streptococci. Likewise, it has been shown that levels of Vibrio cholerae O1 in oysters were not related to either fecal coliform levels in either oysters or ambient waters. Furthermore, the survival of these bacterial indicators (e.g., E. coli and fecal coliform) in the marine environment is very different from that of other pathogens. The above evidence clearly demonstrates that the use of E. coli or fecal coliforms as indicators for waterborne pathogen is largely inadequate. Beril et al. [9] pointed out that the present use of bacteria indicators for sanitary standards is largely unsatisfactory and cannot offer adequate protection to public health.

2.7 Indicators and Evaluation of Health Risk

Indicators are designed to:

(i) Educate/influence policy and decision on status and trends;
(ii) Improve data accessibility;
(iii) Measure success;
(iv) Foster "adaptive management", watershed level ecological health assessment, and support for additional remedial/preventive management;
(v) Promote stewardship through public outreach.

Well-designed and carefully selected indicators serve as practical, economical, and responsive tools for tracking ecosystem changes. Indicators are specific, well-defined, and measurable variables that reflect some key characteristics that can be tracked through time to signal what is happening within and across ecosystems. Indicator categories important include:

- Biological—which includes the indicator harmful algal events, and
- Chemical—which includes the indicator nutrients nitrogen and phosphorus.

While a specific list of the HAB species in the Great Lakes has not been done, it seems they could include most of the ones in the toxigenic cyanobacteria key in Chapman. Ideally, the number of HABs should be quantified on some basis, i.e., low, medium, and high intensity for marine coastal waters and freshwater bodies. However, there is no continent or nationwide monitoring or reporting program for HABs, which are there generally accepted definitions of low, medium, and high density. In addition, there are important indicators for HAB health risk. These would be genera and species of HABs plus the toxins and levels produced.

Since most countries have no national regulations defining acceptable levels of cyanotoxins in drinking or recreational waters, most rely on guidelines published by the WHO or on risk assessments based on the WHO data to manage these water bodies. However, these guidelines were based on cell concentrations rather than on cyanotoxin concentrations, and not all cyanobacterial blooms produce toxins. Resource managers and public health officials are often left with the difficult choice between protecting public health by closing a water body with a significant algal bloom that may not be producing toxins or protecting the local tourism economy by keeping a water body with a visible, aesthetically unappealing bloom open for use. In addition to using the WHO guidance, some states have done their risk assessments to develop guidelines to support public health decision-making, such as posting advisories or closing water bodies. For example, the Office of Environmental Health Hazard Assessment of the California Environmental Protection Agency issued guidance on six cyanotoxins in 2012 [4]. The guidance provides calculated action levels that may be applied by local, regional, state, or tribal entities to reduce or eliminate exposure of people and animals to algal toxins. The legislation also formalizes the responsibility for warning people about cyanobacteria and cyanoHABs blooms and sets the health-related warning thresholds: tourism officials will warn lake users if algae cell counts exceed 100,000 cells/mL and microcystin concentrations exceed 20 μg/L [4]. Since legislation is established on algae cell counts and microcystin concentrations, either of these measures could be used as an indicator of pathogen risk.

Due to these programs and legislation, we are slowly becoming better able to assess risk for HABs now that the appropriate indicators are being defined. This makes it possible to employ a risk assessment matrix for HABs. For example, using the WHO action level of 100,000 cells/ml (for a toxigenic species), blooms exceeding this level should have their source water closed to recreational use until cell levels fall below this level. In a more formal risk assessment, it is also important to know if the toxigenic bloom is producing one or more cyanotoxins, and what the concentrations are. It is now established that blooms can have highly variable levels of cyanotoxins, ranging from non-detect to thousands of micrograms per liter. Most countries set up their initial monitoring and management plans to respond to cell counts once a toxigenic species is identified. This has worked well for the initial establishment of occurrence and distribution of HAB events. However, over time it has become clear that there are an increasing number of events and that it may be too conservative to place warnings on closures for all these events based upon cell counts. The economic costs involved from recreational, fishing, and increased water treatment may be unwarranted in some cases.

Water reservoirs are man-made ecosystems of scientific interest, being the combination of natural phenomena and human manipulation. These ecosystems exist because of their uses and the relative goods and services that they assure to humans (e.g., they are sources for drinking and irrigation water, energy production, aquaculture exploitation, flood management, touristic and leisure activities). Reservoirs have rapidly increased in number and size in the last century, mainly in response to increased water demands for irrigation and energy [42]. In the Mediterranean climate, which dominates regions from 32 to 40 north and south of the Equator and is one of the five major climatic regions, water is becoming the most limiting natural resource due to climate change. In particular, the Mediterranean basin is considered one of the most vulnerable areas of the world for the impacts of warming and modification of precipitation regime. Consequently, further water restriction waits. While attempting to maximize economic profit from water resources exploitation, strong management conflicts are arising among the different stakeholders and institutions for management of reservoirs uses.

Defining the environmental characteristics of reservoirs, their seasonal variations, and the factors that influence phytoplankton growth and cyanobacteria proliferation are indispensable steps for planning both water resource management and, if necessary, ecological quality improvement, in compliance with the Water Framework Directive's requirement of reservoirs being in good condition by 2027 [61]. Moreover, climate change is expected to further increase eutrophication, cyano-HABs and their impacts worldwide, including areas at different latitudes than the Mediterranean climate, raising new challenges for their mitigation and management [56]. However, with the current levels of knowledge, researchers cannot comprehensively evaluate the complex interactions between eutrophication, HABs, and climate change, and filling these knowledge gaps is, nowadays, of paramount importance.

2.8 Prevention

The causes that drive eutrophication are multiple and the mechanisms involved are complex. Several elements should be considered to assess the possible actions aimed at counteracting nutrient enrichment of water supplies. The use of computerized models now allows a better understanding of the role of each factor and predicting the efficiency of various curative and preventive measures. The best way to avoid eutrophication is to try to control those mechanisms that are under human control; this clearly means to reduce the input of nutrients into the water basins. Such control unfortunately does not have a linear effect on the eutrophication intensity.

Integrated management should comprise:

- *Identification of all nutrient sources.* Such information can be acquired by studies of the catchment area of the water supply. Knowledge of industrial activities, discharge practices, and localization, as well as agricultural practices (fertilizer contribution/plant use and localization of crops) is necessary to plan and implement actions aiming at limiting the nutrient enrichment of water. The identification of sewage discharge points, agricultural practices, the nature of the soil, the vegetation, and the interaction between the soil and the water can be of great help in knowing which areas should be targeted.
- *Knowledge of the hydrodynamics of the water body* particularly the way nutrients are transported, and of the vulnerability of the aquifer will allow determination of the ways by which the water is enriched with nutrients.
- *Anthropogenic nutrient point sources* such as nontreated industrial and domestic wastewater discharge can be minimized by the systematic use of wastewater treatments. In sensitive areas, industries, and local authorities should control the level

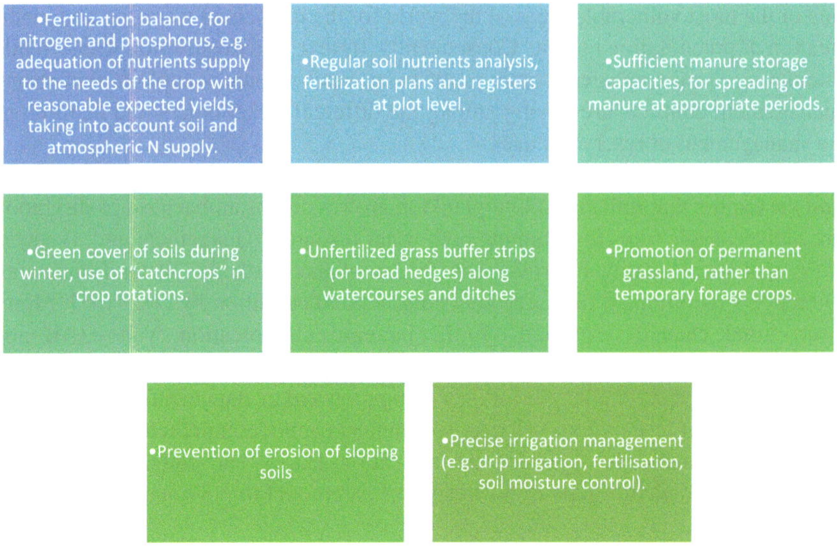

Fig. 2.7 Nutrient loss prevention at farm level and good practices

of nutrients in the treated wastewater using specific denitrification or phosphorus removal treatments. Diffused anthropogenic nutrient sources can be controlled by soil conservation techniques and fertilizer restrictions.

- *Knowledge of the agronomic balance* (ratio of fertilizer contribution to plant use) is very relevant to optimize the fertilization practice and to limit the loss of nutrients.

Diffused nutrient losses could be reduced by implementation at farm level of good practices such as (Fig. 2.7):

In coastal areas, improvement in the dispersion of nutrients, either through the multiplication of discharge points or through the changing of their localization, can help to avoid localized high levels of nutrients. Reuse and recycling, in aquaculture and agriculture, of waters rich in nutrients can be optimized to avoid discharge into the water body and direct consumption of the nutrients by the local flora and fauna.

2.9 European Directives and Guidelines for Prevention

In addition to eutrophication control measures, the European Commission has adopted legislation, which applies to all Member States in the European Union, for the control of eutrophication either directly or indirectly at a national level. The relevant Directives, which all aim to limit pollution, are described below:

a. *The Urban WasteWater Treatment Directive*

This Directive sets minimum standards for the provision of sewage collection and treatment. Discharges of greater than 10,000 population equivalent to waters identified by Member States as sensitive to eutrophication—termed "Sensitive Areas (Eutrophic)"—must comply with specified standards for phosphorus and/or nitrogen (depending on the situation), in addition to a requirement for secondary sewage treatment.

b. *The Nitrate Directive*

This Directive has an objective to "reduce water pollution by nitrates from agricultural sources, and to prevent further such pollution." A water body is identified as polluted if it shows signs of eutrophication—induced by excess nitrates—or if it is used for drinking water abstraction and has elevated concentrations of nitrate.

c. *The Habitats Directive*

This Directive enables the designation of Special Areas of Conservation (SACs) in which rare and endangered habitats and species must be protected. If eutrophication threatens the conservation status of such areas there may be a need to impose nutrient control requirements on discharges, where appropriate. The environment regulators have to review discharge consents to ensure that the designated sites are adequately protected.

d. *The Wild Birds Directive*

This Directive gives the same protection against discharges to sites designated as Special Protection Areas (SPAs).

e. *Water Framework Directive*

The Water Framework Directive incorporates the requirements of the existing Directives, supplemented by whatever additional measures are needed to control pollution sources, both point and diffuse, to achieve the stated objective. It is therefore an essential vehicle for the control of eutrophication, providing a coordinated approach to sustainable water management through river-basin management plans, assessing the state of waters and the sources of pollution, then formulating and implementing improvement plans. The objective is to achieve good ecological quality, which will include avoidance of the effects of eutrophication.

In addition to the EC Directives, there is a range of international commitments, elements of which are of direct relevance to eutrophication control.

The Earth Summit Conference in Rio de Janeiro in 1992 resulted in the UK entering commitments to safeguard the well-being of the planet. Action Plans for priority habitats and species are being developed including those to protect mesotrophic lakes and eutrophic standing waters where control of eutrophication will be important in maintaining or restoring biodiversity.

The identification and control of diffuse agricultural sources of nutrients present the greatest challenges. There are increasing pressures on the agricultural industry to develop more environmentally sustainable farming methods. In several countries, this is being implemented through agri-environment schemes where grants are offered for farming in a more environmentally friendly manner.

References

1. Adams WJ, Cardwell AS, DeForest DK, Gensemer RW, Santore RC, Wang N, Nordheim E (2018) Aluminum bioavailability and toxicity to aquatic organisms: introduction to the special section. 37(1):34–35
2. Arend KK, Beletsky D, Depinto JV, Ludsin SA, Roberts JJ, Rucinski DK, Scavia D, Schwab DJ, Hook T (2011) Seasonal and interannual effects of hypoxia on fish habitat quality in central Lake Erie. Freshw Biol 56:366–383. https://doi.org/10.1111/j.1365-2427.2010.02504
3. Arhonditsis G, Eleftheriadou M, Karydis M, Tsirtsis G (2003) Eutrophication risk assessment in coastal embayments using simple statistical models. Mar Pollut Bull 46:1174–1178
4. Backer L (2012) Freshwater Algal Blooms & Public Health Lake Line. 32(3):7–9
5. Baroni IR, Palastanga V, Slomp CP (2020) Enhanced organic carbon burial in sediments of oxygen minimum zones upon ocean deoxygenation. Front Mar Sci 6:839
6. Belin C, Soudant D, Amzil Z (2020) Three decades of data on phytoplankton and phycotoxins on the French coast: lessons from REPHY and REPHYTOX, 2020 Harmful Algae Available online 26 March 2020, 101733. In Press. https://doi.org/10.1016/j.hal.2019.101733
7. Benndorf J (1990) Conditions for effective biomanipulation - conclusions derived from whole-lake experiments in Europe. Hydrobiologia 200:187–203
8. Berdalet E, Fleming LE, Gowen R, Davidson K, Hess P, Backer LC, Moore SK, Hoagland P, Enevoldsen H (2016) Marine harmful algal blooms, human health and wellbeing: challenges and opportunities in the 21st century. J Mar Biol Assoc UK 96:61–91
9. Beril C, Crance JM, Leguyader F, Apaire-Marchais V, Leveque, Albert M, Goraguer MA, Schwartzbrod L, Billaudel S (1996) Study of viral and bacterial indicators in cockles and mussels. Mar Pollut Bull 32(5):404–409
10. Bird PD, Kraa E (1992) Overview of the 1990 viral gastro-enteritis outbreak from oysters Conference Internationale sur la Purification des Coquillages, Rennes (France), 6–8 Apr 1992
11. Borja VM, Furio EF, Gatdula NC, Iwataki M (2019) Occurrence of harmful algal blooms caused by various phytoplankton species in the last three decades in Manila Bay, Philippines. Phil J Nat Sci 24:80–90
12. Boyd CE, Tucker CS (1998) Pond aquaculture water quality management. Kluwer, Norwell MA
13. Breitburg D, Gregoire M, Isensee K (eds) (2018) The ocean is losing its breath: Declining oxygen in the world's ocean and coastal waters. IOC-UNESCO, IOC Technical Series, No. 137 40 pp (IOC/2018/TS/137)
14. Bricker SB, Longstaff B, Dennison W, Jones A, Boicourt K, Wicks C, Woerner J (2008) Effects of nutrient enrichment in the nation's estuaries: a decade of change. Harmful Algae 21–32
15. Burkholder JM et al (1992) New 'phantom' dinoflagellate is the causative agent of major estuarine fish kills. Nature 358:407–410
16. Carlson RE (1977) A trophic state index for lakes. Limnol Oceanogr 22:361–369
17. Carpenter SR (1981) Submersed vegetation: an internal factor in lake ecosystem succession. Am Nat 118:372–383
18. Carpenter SR et al (1995) Biological control of eutrophication in lakes. Environ Sci Technol 29:784–786
19. Carpenter SR et al (1998) Nonpoint pollution of surface waters with phosphorus and nitrogen. Ecol Appl 8:559–568
20. Chapelle A, Lazure P, Menesguen A (1994) Modelling eutrophication events in a coastal ecosystem. Sensit Anal Estuar Coast Shelf Sci 39(6):529–548

21. Chen et al (2009) Nonproteolytic functions of ubiquitin in cell signalling. Review Mol Cell 33(3):275–286. https://doi.org/10.1016/j.molcel.2009.01.014
22. Chorus I, Bartram J (eds) (1999) Toxic cyanobacteria in water: a guide to their public health consequences, monitoring, and management. E & FN Spon, London UK
23. Condon RH, Duarte CM, Pitt KA, Robinson KL, Lucas CH, Sutherland KR, Mianzan HW, Bogeberg M, Purcell JE, Decker MB, Uye S-I, Madin LP, Brodeur RD, Haddock SHD, Malej A, Parry GD, Eriksen E, Quiñones J, Acha M, Harvey M, Arthur JM, Graham WM (2013) Recurrent jellyfish blooms are a consequence of global oscillations. PNAS 110:1000–1005
24. Crews JR, Chappell JA (2007) Agriculture and natural resources U.S. catfish industry outlook. Auburn University, Auburn AL
25. Diaz RJ, Rosenberg R (2008) Spreading dead zones and consequences for marine ecosystems. Science 321:926–929
26. Dittmann E, Fewer DP, Neilan BA (2013) Cyanobacterial toxins: biosynthetic routes and evolutionary roots. FEMS Microbiol Rev 37:23–43
27. Dodds WK, Bouska WW, Eitzmann JL, Pilger TJ, Pitts KL, Riley AJ, Schloesser JT, Thornbrugh DJ (2009) Eutrophication of U.S. freshwaters: analysis of potential economic damages. Environ Sci Technol 43:12–19
28. Downing JA et al (2001) Predicting cyanobacteria dominance in lakes. Can J Fish Aquat Sci 58:1905–1908
29. Edmondson WT (1970) Phosphorus, nitrogen, and algae in Lake Washington after diversion of sewage. Science 169:690–691
30. Estrada V, Parodi ER, Diaz MS (2009) Determination of biogeomechenical parameters in eutrophication models with simultaneous dynamic optimization approaches. Comput Chem Eng 33(10):1760–1769
31. Eutrophication and Health. European Commission (2002) Office for official publications of the European Communities: Luxembourg. ISBN 92-894-4413-4.28 pp
32. Farley (2012) Eutrophication in fresh waters: an international review, encyclopedia of lakes and reservoirs. 2012 Edition | Editors: Lars Bengtsson, Reginald W. Herschy, Rhodes W. Fairbridge https://doi.org/10.1007/978-1-4020-4410-6_79
33. Ferley JP, Zmirou D, Balducci D, Baleux B, Fera P, Larbaigt G, Jacq E, Moissonnier B, Blineau A, Boudot J (1989) Epidemiological significance of microbiological pollution criteria for river recreational waters, Int J Epidemiol 18(1):198–205. 10.1093/ije/18.1.198
34. Francis G (1878) Poisonous Australian lake. Nature 18:11–12
35. Funkey CP, Conley DJ, Reuss NS, Humborg C, Jilbert T, Slomp CP (2014) Hypoxia sustains cyanobacteria blooms in the Baltic Sea. Environ Sci Technol 48:2598–2602
36. Halliday M, Kang L-Y, Zhou T-K, Hu M-D, Pan Q-C, Fu T-Y, Huang Y-S, Hu S-L (1991) An epidemic of hepatitis A attributable to the ingestion of raw clams in Shanghai, China. J Infect Dis 164(5):852–859
37. http://www.coastalwiki.org/wiki/Possible_consequences_of_eutrophication#Humanhealth_impacts
38. Huisman J et al (2004) Changes in turbulent mixing shift competition for light between phytoplankton species. Ecology 85:2960–2970
39. Hyytiainen K, Hasler B, Ericsdotter S, Nekoro M, Blyh K, Artell J, Ahlvik L, Ahtiainen J (2013) Worth it: benefits outweigh costs in reducing eutrophication in the Baltic. Baltic STERN Summary Report for HELCOM 2013 Ministerial Meeting Arend, KK et al Seasonal and interannual effects of hypoxia on fish habitat quality in central Lake Erie. Freshw Biol 56:366–383
40. Jeppesen E et al (1997) Top-down control in freshwater lakes: the role of nutrient state, submerged macrophytes and water depth. Hydrobiologia 342(343):151–164
41. Johnson PT, Chase JM, Dosch KL, Hartson RB, Gross JA, Larson DJ, Sutherland DR, Carpenter SR (2007) Aquatic eutrophication promotes pathogenic infection in amphibians. Proc Natl Acad Sci USA 104:15781–15786
42. Jørgensen SE, Costanza R, Xu F-L (2005) Handbook of ecological indicators for assessment of ecosystem health. https://doi.org/10.1201/9780203490181

43. Jüttner F, Watson SB (2007) Biochemical and ecological control of geosmin and 2-methylisoborneol in source waters. Appl Environ Microbiol 73:4395–4406. Kouakou CRC, Poder TG (2019) Economic impact of harmful algal blooms on human health: a systematic review. J Water Health 17(4):499–516
44. Lehtiniemi M et al (2005) Turbidity decreases anti-predator behaviour in pike larvae, Esox Lucius. Environ Biol Fishes 73:1–8
45. Li G, Cheng L, Zhu J, Trenberth KE, Mann ME, Abraham JP (2020) Increasing ocean stratification over the past half-century. Nat Clim Change. https://doi.org/10.1038/s41558-020-00918-2
46. Marchi M, Jørgensen SE, Bécares E, Fernández-Aláez C, Rodríguez C, Fernández-Aláez M, Pulselli FM, Marchettini N, Bastianoni S (2012) Effects of eutrophication and exotic crayfish on health status of two Spanish lakes: a joint application of ecological indicators. Ecol Indic 20:92–100
47. Masó M, Garcés E (2006) Harmful microalgae blooms (HAB); problematic and conditions that induce them. Mar Pollut Bull 53(10–12):620–630. https://doi.org/10.1016/j.marpolbul.2006.08.006. Epub 2006 Sep 28
48. Le Moal M, Gascuel-Odoux C, Ménesguen A, Souchon Y, Étrillard C, Levain A, Moatar F, Pannard A, Souchu P, Lefebvre A (2019) Eutrophication: a new wine in an old bottle? Sci Total Environ 651:1–11
49. Morris JG (1999) Harmful algal blooms: an emerging public health problem with possible links to human stress on the environment. Annu Rev Energy Env 24:367–390
50. Morton B (1989) Pollution of the coastal waters of Hong Kong. Mar Pollut Bull 20(7):310–318
51. Nasser AM (1994) Prevalence and fate of hepatitis a virus in water. Crit Rev Environ Sci Technol 24(4)
52. Nixon SW, Buckley BA (2002) "A strikingly rich zone"—Nutrient enrichment and secondary production in coastal marine ecosystems. Estuaries 25:782–796
53. Nobre AM, Ferreira JG, Newton A, Simas T, Icely JD, Neves R (2009) Management of coastal eutrophication: integration of field data, ecosystem-scale simulations and screening models. J Mar Syst 56(3–4):375–390
54. Osborne NJ, Webb PM, Shaw GR (2001) The toxins of Lyngbya majuscula and their human and ecological health effects. Environ Int 27:381–392
55. Paerl HW (1988) Nuisance phytoplankton blooms in coastal, estuarine, and inland waters. Limnol Oceanogr 33:823–847
56. Paerl HW, Huisman J (2009) Climate change: a catalyst for global expansion of harmful cyanobacterial blooms. Environ Microbiol Rep 1:27–37
57. Paerl HW, Paul VJ (2012) Climate change: links to global expansion of harmful cyanobacteria. Water Res 46:1349–1363
58. Parsons GR, Morgan AO, Whitehead JC, Haab TC (2006) The welfare effects of Pfiesteria-related fish kills: a contingent behavior analysis of seafood consumers. Agricult Resou Econom Rev 35(1)
59. Patricia GM, Sybil S, Cynthia HA, Joann BM, Louis CA, Vince K (2005) The role of eutrophication in the global proliferation of harmful algal blooms new perspectives and new approaches. Oceanography 18:198–209
60. Pick FR (2016) Blooming algae: A Canadian perspective on the rise of toxic cyanobacteria. Can J Fish Aquat Sci 73:1149–1158
61. Poikane S, Portielje R, van den Berg M, Phillips G, Brucet S, Carvalho L, Mischke U, Ott I, Soszka H, Van Wichelen J (UGent) (2014) Defining ecologically relevant water quality targets for lakes in Europe. J Appl Ecol 51(3):592–602
62. Porter KG (1977) The plant-animal interface in freshwater ecosystems. Am Sci 65:159–170
63. Pretty JN, Mason CF, Nedwell DB, Hine RE, Leaf S, Dils R (2003) Environmental costs of freshwater eutrophication in England and Wales. Environ Sci Technol 37:201–208
64. Santiago S, Thomas RL, Larbaigt G, Corvi C, Rossel D, Tarradellas J, Gregor DJ, McCarthy L, Vernet JP (1994) Nutrient, heavy metal and organic pollutant composition of suspended and bed sediments in the Rhone River. Aquatic Sci 56:220–242

65. Schindler DW (1974) Eutrophication and recovery in experimental lakes: implications for lake management. Science 174:897–899
66. Schindler DW (2006) Recent advances in the understanding and management of eutrophication. Limnol Oceanogr 51:356–363
67. Schindler DW (2012) The dilemma of controlling cultural eutrophication of lakes. Proc R Soc B Biol Sci 279:4322–4333
68. Selman M, Greenhalgh S, Diaz R, Sugg Z (2008) Eutrophication and hypoxia in coastal areas: a global assessment of the state of knowledge of water quality. Eutrophication Hypoxia 1
69. Shapiro J et al (1975) Biomanipulation: an ecosystem approach to lake restoration. In: Brezonik PL, Fox JL (eds) Water quality management through biological control. University of Florida, Gainesville, FL, pp 85–96
70. Sidharta BR (2005) The current status of research on harmful algal bloom (HAB) in Indonesia. J Coast Developm 8(2):75–88
71. Smetacek V, Zingone A (2013) Green and golden seaweed tides on the rise. Nature 504. 10.1038/nature12860
72. Smith VH, Schindler DW (2009) Eutrophication science: where do we go from here? Trends Ecol Evol 24:201–207
73. Srinivasan R, Sorial GA (2011) Treatment of taste and odor causing compounds 2-methyl isoborneol and geosmin in drinking water: a critical review. J Environ Sci 23:1–13
74. Søndergaard M et al (2003) Role of sediment and internal loading of phosphorus in shallow lakes. Hydrobiologia 506–509:135–145
75. Tillmanns AR et al (2008) Meta-analysis of cyanobacterial effects on zooplankton population growth rate: species-specific responses. Fundam Appl Limnol 171:285–295
76. Turner AM, Chislock MF (2010) Blinded by the stink: nutrient enrichment impairs the perception of predation risk by freshwater snails. Ecol Appl 20:2089–2095
77. Wallace RB, Baumann H, Grear JS, Aller RC, Gobler CJ (2014) Coastal ocean acidification: The other eutrophication problem. Estuar Coast Shelf Sci 148:1–13
78. Watson PH, Barnes RO (2011) A proposed schema for classifying human research biobanks. Published 2011 Biopreservation and Biobanking 9(4)
79. Wells CP, Lavretsky P, Sorenson MD, Peters JL, DaCosta JM, Turnbull S, Uyehara KJ, Malachowski CP, Dugger BD (2019) Persistence of an endangered native duck, feral mallards, and multiple hybrid swarms across the main Hawaiian Islands. 28(24):5203–5216
80. Willén E, Ahlgren G, Tilahun G, Spoof L, RNeffling M, Meriluoto J (2011) Cyanotoxin production in seven Ethiopian Rift Valley lakes. Inland Waters 1:81–91
81. Wilson AE et al (2006) Effects of cyanobacterial toxicity and morphology on the population growth of freshwater zooplankton: meta-analyses of laboratory experiments. Limnol Oceanogr 51:1915–1924
82. Xu ZY, Jian ZHL, Zai XW, De PX, Dong X (1992) Ecology and prevention of a shellfish-associated hepatitis A epidemic in Shanghai. China Vaccine 10(1):S67–S68
83. Zewde TW, Johansen JA, Kifle D, Demissie TB, Hansen JøH, Tadesse Z (2018) Concentrations of microcystins in the muscle and liver tissues of fish species from Koka reservoir, Ethiopia: a potential threat to public health. Toxicon https://doi.org/110.1016/j.toxicon.2018.08.013. Accessed on 20 September 2018

Chapter 3
The Effect of Immediate Treatment for Water Quality: Policies and Protection Perspectives

Grigorios L. Kyriakopoulos

Abstract The restoration of water sources that are eutrophication-burdened is certainly bounded on the legal framework that governs water protection in European and international contexts. In the European Union, the Water Framework Directive (2000/60/EC) is proven of utmost importance legislative tool since it sets specific normative targets and regulatory policies to improve water quality. However, many water ecosystems have not been able to meet the targets set by the global covered regulative Directives to date. The main reason is the long time lag of nature rebound to an acceptable water quality, based on the carrying capacity and the internal phosphorus migration from the sediment. Consequently, in this chapter, the main restoration methods of eutrophic lakes have been examined, being proven critical indicators of (a) reducing the internal sources of phosphorus from the sediment, (b) improving water quality, (c) controlling the water eutrophication levels in the long-run while simultaneously (d) achieving environmental conditions of sustainability and legislative goals through the framework of the EU Water Framework Directive 2000/60/EC.

3.1 Introduction

In the European Union (EU) context, a more comprehensive and protective, against eutrophication, legislation has been issued by conventions and regulations. In particular, surface and groundwater sources have been considered by the EU Water Framework Directive 2000/60/EC (hereinafter it is noted as "EU WFD") [1], while a framework for the marine environmental policy up to the 200 nautical miles limit of the European exclusive economic zone has been developed by the Marine Strategy Framework Directive (MSFD) [2]. Similarly in the US, the legislative framework has been provisioned towards the protection of coastal water quality: Harmful Algal Bloom and Hypoxia Research and Control Amendments Act (HABHRCA) [3].

G. L. Kyriakopoulos (✉)
Photometry Laboratory, Electric Power Division, School of Electrical and Computer Engineering, National Technical University of Athens, 15780 Athens, Greece
e-mail: gregkyr@chemeng.ntua.gr

M. G. Zamparas and G. L. Kyriakopoulos (eds.), *Chemical Lake Restoration*,
https://doi.org/10.1007/978-3-030-76380-0_3

Due to the plethora of natural water sources among the EU and the imperative need to regulate this precious natural wealth, there has been already developed one of the oldest and highly developed water legislation containing ongoing and progressive areas of environmental policy [4]. Among the various EU freshwater environmental policies that of EU WFD is of over-arching importance. Following almost 30 years of EU water legislation regarding individual issues with some considerable progress, the EU WFD was introduced in 2000, signaling a new era of water policy [4].

Despite the important and tenuous efforts invested towards the EU WFD coordination among EU Member States (hereinafter it is noted as "EU MS") and the strict timetables incurred, the implementation progress has been still very challenging to achieve the EU WFD objectives and to improve slow adoption pace to ecological status of waters among EU MS (Fig. 3.1). It is noteworthy that in 2015 nearly half of EU surface waters did not meet a good ecological status, while a 40% portion among EU MS water sources contains an unknown chemical status (European Commission 2012a). Moreover, almost a quarter of all environment-related infringements, counted for 73 non-implementation cases of water legislation among EU MS, are still open [6].

In a similar study, it was stated that a significant part of EU MS water bodies has not met good status objectives that are laid down in the EU WFD, being reported at the European Environment Agency's 2018 State of Water, and the Commission's assessment report of the second River Basin Management Plans (RBMPs). However,

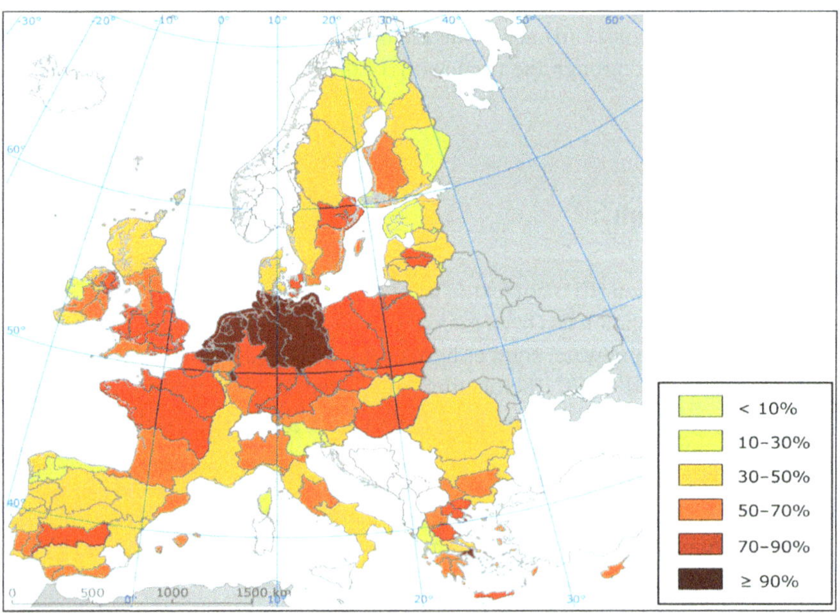

Fig. 3.1 Profile of classifying lake water bodies among various EU River Basin Districts holding disadvantageous (deviant from good) ecological status or potential [5]

the specific objectives had scheduled to be met by 2015, having the tolerance of adaptability and reliance on exemptions if certain conditions are met [7].

It is noted that during the first cycle of RBMPs, progress was made across member states in improving chemical status which indicates that certain effective measures were implemented. When the chemical status of surface water bodies is compared between the first and second set of RBMPs, it is clear that the number of water bodies registered as "unknown" status has dropped significantly. Generally, chemical status has declined in lakes, remained similar in rivers, and improved in transitional/coastal waters. However, despite these improvements, as illustrated in Fig. 3.2, there is still a significant number of no good chemical status of surface water bodies.

Following the EU WFD's objectives, the portions of only 40% for ecological status and that of 38% for chemical status have been achieved for surface water bodies. The situation is better for groundwater. Contrarily, a good chemical status and a good overall status of 84% and 92% of groundwater bodies have been reported, respectively. The aforesaid 84% portion is representing 74% of the total groundwater body area, while the aforesaid 92% portion is representing 89% of the total groundwater body area. Nevertheless, compared to the first cycle, no substantial improvement in water bodies regarding their overall status was disclosed [7, 9].

The main reasons for such procedural constraints in which no substantial progress at the overall water bodies have been achieved between the two cycles reside in the fact that the second RBMPs (due in 2015) expressed a general show classification status by the period of 2012–2013. By that time many measures were still in the implementation process. Among such slow pace of implementation, there were cases of slow time in nature's response, thus, taking extensive time before measures to be effected on. Another explanation is attributed to increased monitoring, where some water bodies previously had been classified as "unknown" and they are now characterized as "less than good". Moreover, in the second cycle, there is the frequent utility of overall stricter methodologies due to the harmonization of methodologies [10]. Besides, there is an affectionate role of the one-out-all-out principle where, regardless of the quality status of other elements, if the status of one quality element is less than good, then, the overall status of the water body will also be less than good [11]. Such a principle is an essential and ecologically protective tool of integrated water management, and it is difficult to monitor the progress in a water body since the status of that water body will not change until all other problems are addressed [12].

The next target of European water policy for rivers, lakes, coastal and traditional waters by 2027 is the achievement of a good ecological status. More than half of these water bodies are currently under depletion conditions, suffering from the main culprit of nutrients' enrichment. Precisely, a rough estimation of 57% of rivers, 44% of lakes, 40% of coastal waters and 66% of transitional waters are currently behind the good ecological status achievement [13].

Regarding the criteria of EU WFD there are two characteristic cases: (a) shallow eutrophic lakes are mainly confronting difficulties of criteria meeting without restoration and (b) in the UK, the examination of the total sample of more than 14,000 lakes of less than 1 ha in size, a 51% has been likely to require P-decrease measures to meet the attribution of "good status" in terms of EU WFD [14]. Consequently, the adoption

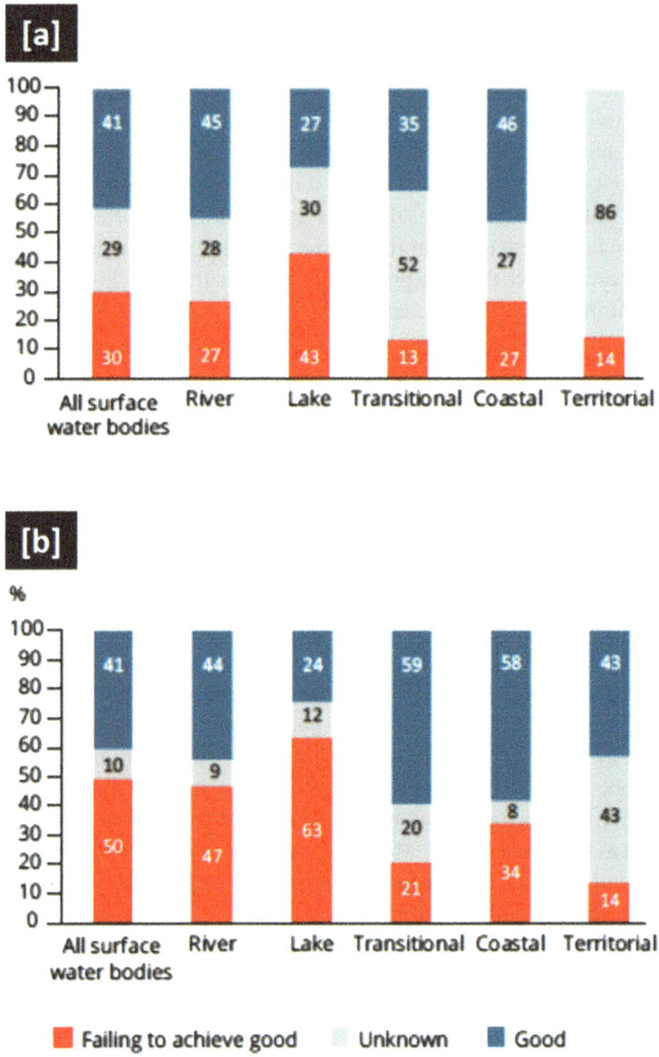

Fig. 3.2 Profile of water categories and corresponding changes' profile in the chemical status of surface water bodies [8]

of regional water action plans should support lakes to achieve the ecological conditions of "moderate", "high", or "good" ecological conditions [15]. Subsequently, there is an imperative need for the establishment of those reliable and comparable nutrient criteria that should be related to good ecological status [16] (Fig. 3.3).

According to Solheim et al. [15], the measurable evaluation of policies undertaken to reduce the environmental burden on lakes and rivers remains a challenging endeavour. In this respect, the managerial pressure of environmental burden can be

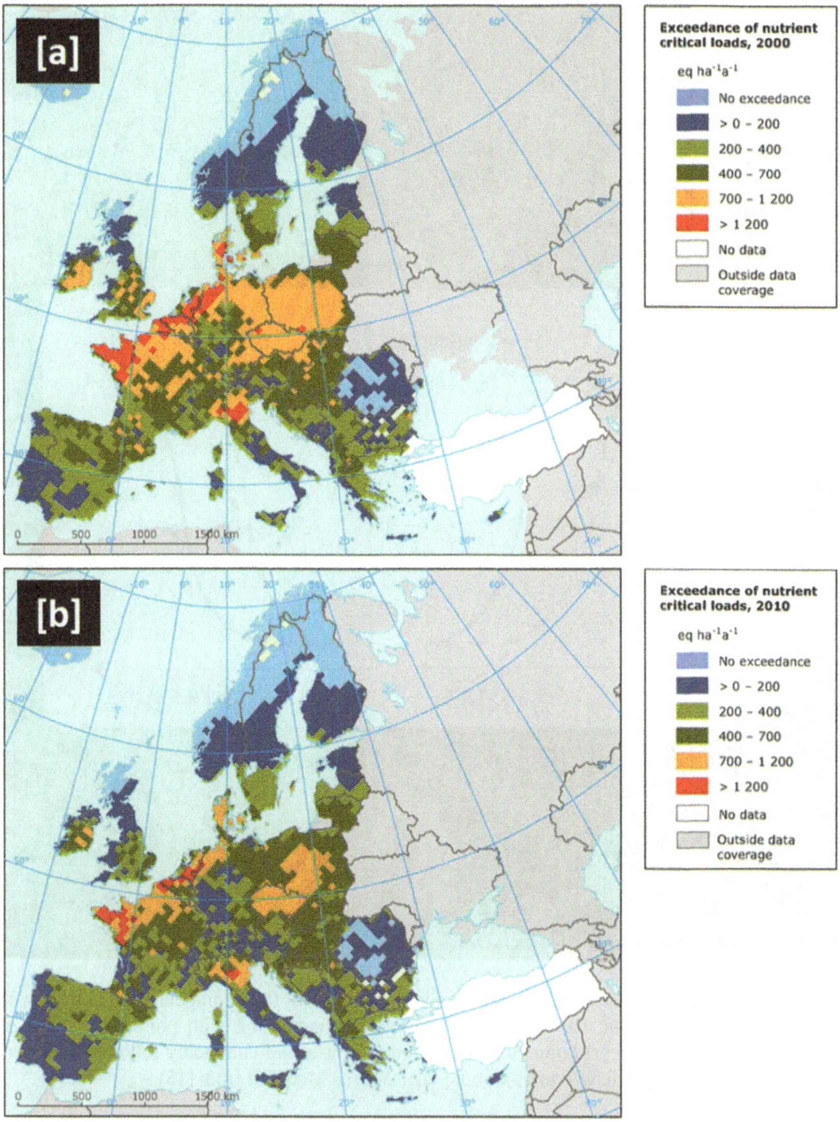

Fig. 3.3 Excessive amount of critical loads for eutrophication attributed to N-nutrient deposition at the years: **a** 2000 and **b** 2010 [17]

attributed to the response curves for various biological elements. Due to hysteresis and resilience that causes long delays in the lake response, response curves of pressure reduction can be differentiated to curves of pressure increase. Consequently, the system rebound to its initial status could be impossible, unless the pressure reduction way below the original threshold (Fig. 3.4).

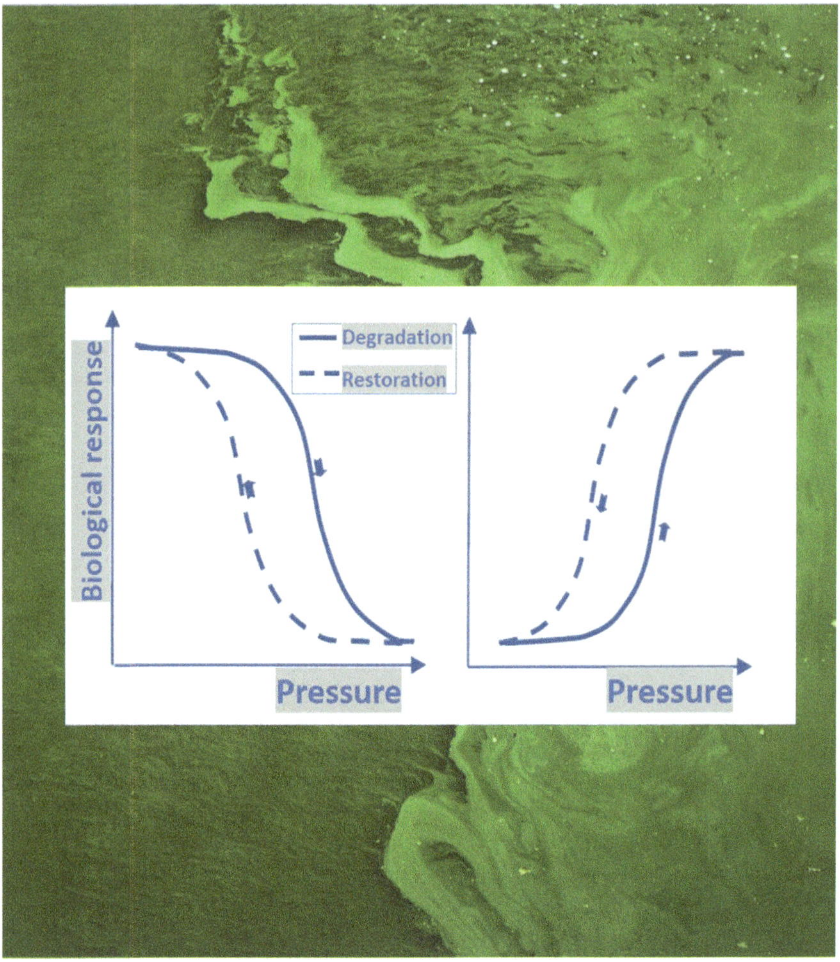

Fig. 3.4 Generalized biological threshold responses to pressure for a component decreasing with pressure (left, e.g., benthic fauna diversity vs. pH) and increasing with pressure (right, e.g., Cyanobacteria vs. TP). Degradation and restoration responses are commonly following different trajectories due to hysteresis and to other time-lags. *Source* Modified from [15]

The impact of WFD on health conditions recovery of water bodies has been oriented to reduce nutrient loads to lakes and reservoirs. The deployment of management actions at the catchment level aims at reducing or eliminating external sources of nutrients, while other lakes and reservoirs are still failing to meet the water quality requirement, showing nutrients' excess that is historically accumulated in waters and sediments, failing to address such severe environmental problems [18–21].

The in-situ and point-source nutrient pollution reduction cannot guarantee per-se the eutrophication lowering and control. In such a case, the Netherlands is one of the few EU countries that is fully compliant to both tertiary urban wastewater treatment for nutrients' elimination, as well as to the Urban Waste Water Treatment Directive [22]. In this case, the water quality in the Netherlands is characterized by an average of 84.5% N-nutrient and 86.8% P-nutrient removal efficiency in 2016, respectively [22]. Despite the fact that more than 99% of the wastewater has undergone treatment, the Netherlands has shown one of the poorest surface water quality among EU MS, where moderate ecological status is preserved by 40% of local lakes (n = 179), while poor or bad ecological status is preserved by 60% of local lakes (n = 269) [13]. Moreover, even when implementing all EU WFD measures, it is anticipated that by 2027 about the moderate portion of 40% of Dutch water bodies could meet the EU WFD goals [22]. This constraint is attributed to the continuing occurrence of sewage overflows and diffuse pollution that is severely impacting water quality [22, 23].

In a similar study, diffuse pollution—which is mostly generated from massive agricultural industry—is mainly exerting pressure on the Dutch surface waters [24]. On the other hand, the Netherlands is representing an export value of 90.3 billion euros in 2018 and 94.5 billion euros in 2019, being considered among global leaders, as the world's second-largest exporter of agricultural goods [22]. In particular, three-quarters of Dutch meat is being exported, making the national meat sector a main source of nutrients' load. Indeed, the N-surplus in 2017 was counted at 319 million kg; a portion of this surplus is soil and atmosphere losses, which is 225 and 94 million kg, respectively. Besides, the P-surplus that ended up in soils was counted as 4 million kg. The effluent of nutrients' leaching from soil to surface waters was 44.6 million kg N, or equivalent to 58% of total load, and 3.95 million kg P, or equivalent to 62% of the total load. The effluent of such huge sources and quantities diffused is implying that among within-system interventions in the Netherlands, they are inevitable in order to accomplish the social auspices of quality demands for water recreation purposes, while simultaneously abiding by national or EU legislation, such as that of EU WFD [22]. An indicative profile of estimating N-total inputs from Denmark to the Danish Straits and a diagrammatic representation of ecosystem state changes of bottom water O_2 concentrations in terms of nutrient concentrations are indicatively depicted in Fig. 3.5.

3.2 History Background of the European Water Legislation

Dating back to the first UN conference on the environment it has been declared an ongoing EU environmental policy from 1973 onwards, flanking economic expansion and called for an action programme by the European Council. Subsequently, instrumental guidance of the progression of EU environmental policy has been remarkably practiced by periodical issues of Community Environmental Action Programs (EAP) that are still now developing and updating [27–29].

The EU considerations of setting common European standards at rivers' and lakes' water for uses at drinking and bathing have reflected the European water legislation dating back to 1975 [7]. Later in 1980, there were issued binding targets for drinking water quality and, consequently, relevant legislation was produced for

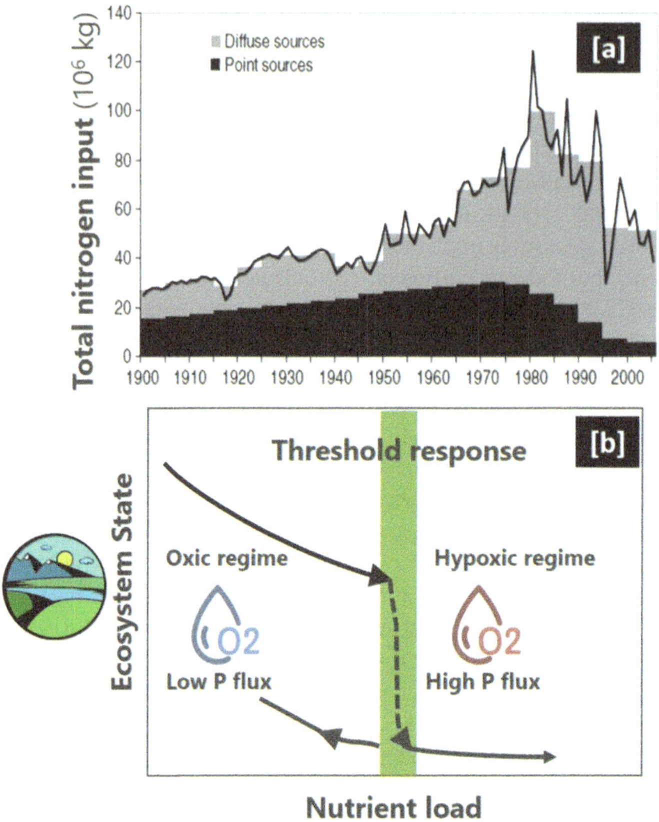

Fig. 3.5 **a** Profile of estimating N-total inputs (solid line) from Denmark to the Danish Straits including the Kattegat since 1900, with averages of five years from point and diffuse sources. *Source* Modified from [25]. **b** Diagrammatic representation of ecosystem state changes of bottom water O_2 concentrations in terms of nutrient concentrations. Killing bottom-living organisms enables the system to shift to anaerobic-dominated processes, then the ecosystem is reaching a threshold. *Source* Modified from [26]

the quality of fish waters, shellfish waters and groundwater (Fig. 3.6). By that time the main applicable emission-control instrument of water-related directives was that of Dangerous Substances Directive [30].

Thereafter, in 2000, consolidation efforts were made for the EU water policies, resulting in the EU WFD adoption. The EU WFD promoted a more holistic approach to water policy, streamlining both the existing freshwater legislation and approaching a river basin management [1]. It can be signified by the EU WFD provision that this Directive can complement to fully refine the water status assessment. In this respect the other legislative tools, Directives, in the field of water policy: Environmental Quality Standards Directive (EQSD, 2008/105/EC) and Groundwater Directive (GWD, 2006/118/EC) were developed in 2008 and 2006, respectively (Fig. 3.7) [32].

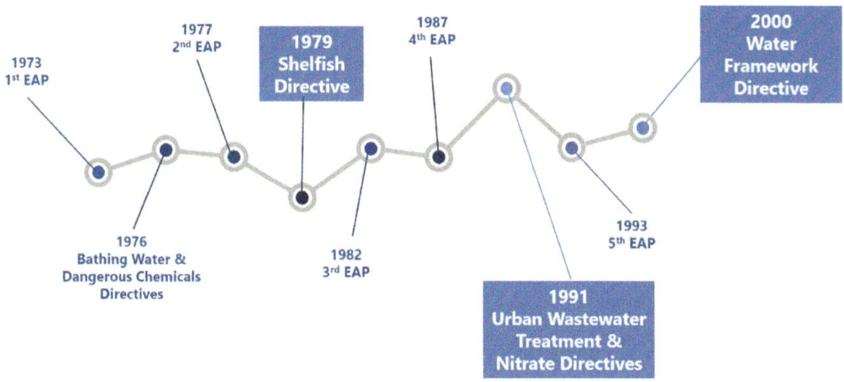

Fig. 3.6 Water policy evolution towards the EU WFD (1973–2000). *Source* Modified from [31]

Fig. 3.7 Historical timeline corpus of WFD implementation milestones, and the relevant policy developments adopted for the period 2000–2027. *Source* Modified from [31]

Similarly, in 2007, the issued Floods Directive consisted a comprehensive and integrated tool for floods management after a devastating and destructive flooding period, 1998–2004, of almost 100 major flood incidents occurred in Europe; among them are the striking August 2002 flood on the Danube, the Elbe and their tributaries [33].

The EU WFD sustains fresh, coastal, and transitional waters ensuring an integrated water management of whole ecosystems. Besides, the EU WFD can be linked in a synergistic manner to several other EU directives that are relevant to water management, sustaining the following environmental objectives [1]:

- Prevention of the status of water bodies from deterioration and depletion;
- Protection, enhancement and restoration of all water bodies, ensuring the good ecological status, or the good ecological potential, in the light of the good chemical

status for surface waters; as well as keeping good quantity and good chemical conditions of groundwater sources by 2015 (Article 4(1)).

Key proactive environmental objectives of deterioration prevention can fit into those general purposes of the EU WFD (being stated in Article 1) ensuring the protection of various water sources—e.g., inland surface, transitional, coastal and groundwater—through the promotion of sustainable water use and the better protection of the aquatic environment.

3.3 Innovative Approach to Water Management

The holistically-EU WFD can manage and protect aquatic ecosystems without necessarily focusing on specific aspects of water quality. Such pictorial representation of the linkages developed among all uses and users of water involved is denoted in Fig. 3.8.

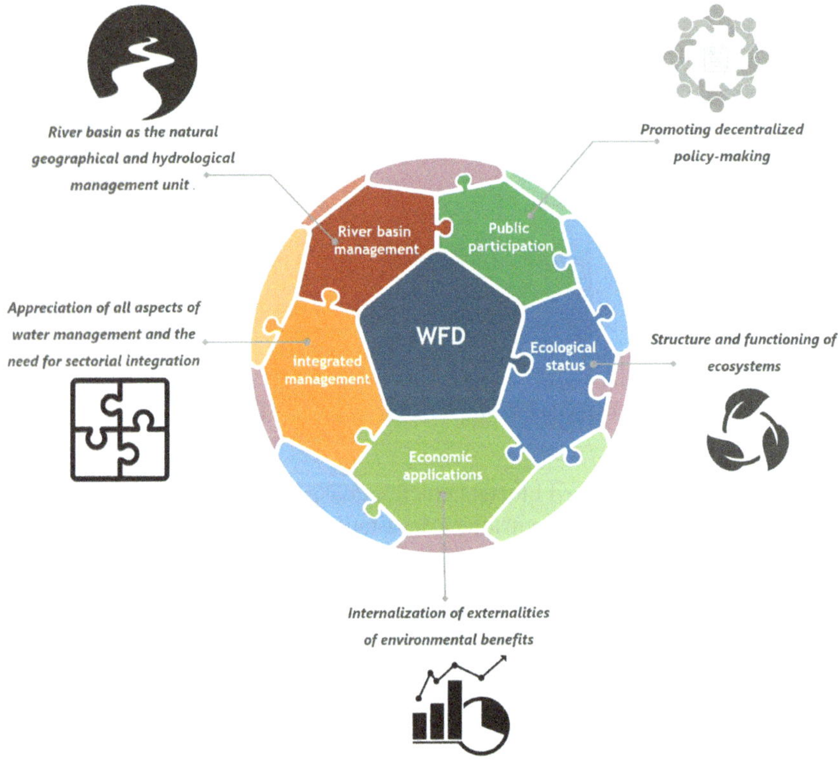

Fig. 3.8 Basic innovations introduced by the EU WFD. *Source* Author's own study

It is noteworthy that while traditional managerial approaches are focused on pollution control, the EU WFD is ensuring ecosystem integrity, being reflected on the "one-out-all-out" principle. The "one-out-all-out" principle at groundwater body implies that if one of either the chemical-status or quantitative-status, determining parameters is less than good, then, the status of the water body will not be classified as good [8]. In the case of surface water, it is of utmost importance noting that if determining the ecological or chemical status as "less than good", then, the status of the water body will not be classified as good. Therefore, based on the EU WDF principle, it is not just sufficient the good chemical status (below the set threshold's value) of prioritizing substances in the surface water, but it is also needed the water body to be in good ecological status. Such water body requirements can be met and defined by the presence of, indicatively, phytoplankton (micro-algae), macrophytes (water plants), benthic invertebrates or fish.

Except for reasoning exemption cases, EU MS are normally called to ensure and protect a "good status" for surface- and ground-water bodies. Targets' time extensions from 2021 up to 2027, the latest, are provisioning only in cases of "good status" to be achieved. In particular, under the EU WFD principles, the achievement of "good status" entails the accomplishment standards of ecology, chemistry and quantity of waters. A "good status" of water sources is implying that water characteristics are very close to the standards of undisturbed water sources, or tolerating only slight changes and deviations from the aforesaid standards of water quality [6]. In this context, there is also a proactive measure of a general "no deterioration" marking, thus, enabling the degradation of water standards and overall physico-chemical status' concerned. The panorama of surface- and ground-water sources' profile comprehended by the EU WFD framework is depicted in Fig. 3.9.

3.4 Strengths and Weaknesses of the Water Framework Directive

The EU WFD (2000/60/EC) is considered the cornerstone of EU MS water policy that applies to all surface waters (rivers, lakes, transitional and coastal waters) as well as groundwater. The immense value and the reasonable water management responsibility are clearly targeted through EU WFD, appreciating that "*water is not a commercial product like any other but, rather, a heritage which must be protected, defended and treated as such*" [4].

Within EU MS, the EU WFD has been adopted to succeed and replace traditional management practices:

- Focusing on understanding and integrating all aspects of the water environment to be effective and sustainable [4].
- Predicated to command and control paradigms looking at pressures in isolation and reduced environmental systems to their constituent elements when setting specific water objectives [34].

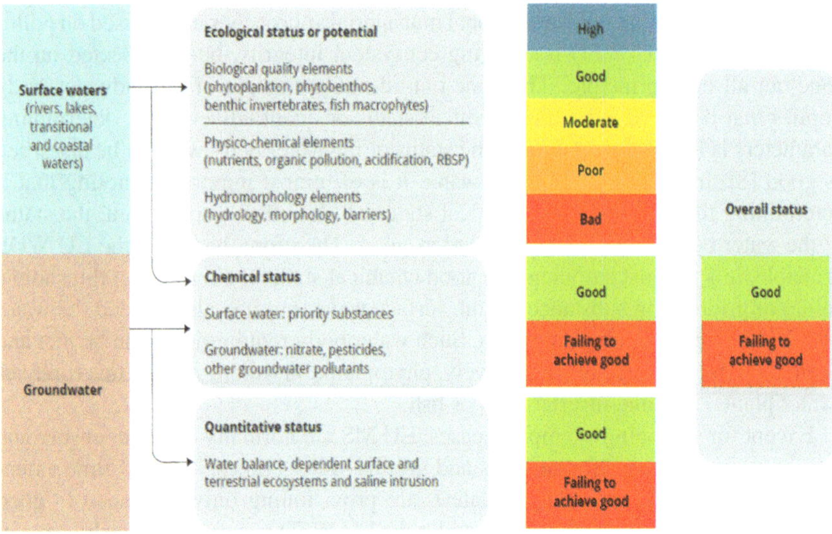

Fig. 3.9 Surface water and groundwater assessment status based on the EU WFD [8]

- Monitoring specific parameters at the point of discharge to control the emissions of individual pollutants beyond specified limits [35].

Considering that individual management of the non-compliant elements could support the complete improvement in ecosystem health [35, 36], EU WFD is a discipline focused on and driven to comprise isolated components of an environmental system, thus, fostering their predictability and stability [34]. While this paradigm shows a long-term effectiveness, enabling industrial western societies to confront severe health-threatening environmental episodes, it is proven ineffective to address the scalability and complexity prospects of ecosystems or their interactive trade-offs, accordingly [34].

The introduction of the EU WFD is primarily aiming to envisage fragmented policies towards envisaging a holistic approach while (a) recognizing the multiple or combative uses of water and (b) increasing awareness of the linking water-, physical- and socio-economics-systems of EU WFD shaping. Considering that catchments have inherently diversified socio-political and natural conditions, institutional shifts of both catchment management and systemic thinking are anticipated. In such a systemic approach, it is required a better understanding of the various socio-economic drivers between land- and water-management [37], including the integration in freshwater governance and impetus for the creation of decentralized policies.

While envisaging the EU WFD Common Implementation Strategy (CIS), being considered a recursive process regarding provisional goal-setting and learning-by-revision [38, 39], the EU WFD is offering an experimentation approach to water governance which advocates EU MS choices in a flexible manner, comparing to earlier Directives policy learning opportunities [38, 39]. While earlier environmental Directives are presuming specific targets, the EU WFD is conceptualized as a no target-based legislative tool, except for the EU WFD's explicit obligation that no water bodies be prone to status deterioration from one class to another. The EU WFD is setting specific operational and technical obligations of implementation that could be resolved by the EU Court of Justice if not followed correctly by the EU MS. The EU WFD is appreciated as a European Directive on environmental and sustainability impacting, and due to this point of view, the introduction of its revolutionary innovation can enhance the prestige of the Directive and its structural adaptability for future environmental regulations [38–40].

The beneficial performance of EU WFD has been already reported in the relevant literature since, during the first EU WFD cycle operated for the period 2009–2015, the number of surface water bodies in "good status" steadily increased by 10% [41]. This beneficial outcome, in alignment with continuous rounds of reviewing and highlighting drawbacks and weaknesses, has proven the tangible impact of the EU WFD effectiveness, as a capable policy tool towards the protection and the regulation for water sources [41, 42].

3.5 Water Framework Directive and Eutrophication

European policy has consistently identified eutrophication as a priority issue for water protection. Substantial progress has been made in combating eutrophication, but there are still various issues where coordination is needed in order to harmonized results for differentiated policy areas to be achieved, such as:

- Harmonizing of assessment methodologies and joint approved criteria of eutrophication, mainly elements and their parametric indicators for rivers, lakes, transitional, coastal and marine waters;
- Utilizing those water-typed specific objectives of biological and general physico-chemical interests in the containing elements;
- Co-ordination of monitoring and reporting;
- Harmonizing of models in order to assess or to predict anthropocentric loading of natural nutrients into inland and marine waters, being based on information about nutrient sources' fate and route-based scenarios (e.g., Harmonized Procedures for Quantification of Catchment Scale Nutrient Losses from European Catchments: "EUROHARP" models);
- Systematic identification of nutrients' sources, acknowledging and co-evaluating possible restoration measures suiting to water bodies.

3.5.1 Nutrient Criteria for Surface Waters Under the Water Framework Directive

While the European water policy is directed to accomplish good ecological status among all water types of local interest by 2027, more than half of water types are in a degraded condition, being burdened by nutrient surplus. Thus, there is an imperative need for reliable and comparable establishment of nutrients' profile that is characterized of good ecological status [16].

European water policy aims to attain while such a good ecological status is defined as closely reflecting real-natural conditions of all water types [1]. There are substantial distortions and deviation of such a "good status" target [8]: 57% of rivers, 44% of lakes, 40% of coastal waters and 66% of transitional waters. An almost 50% deviation of such "good status" among all EU MS water types is attributed to more than one of the following pressures of human activities, including agriculture, urbanization, as well as to hydropower generation and climate change. Besides, only 18% of surface water types having no identifiable significant pressures [13]. The main source of diffuse and point-sourcing of nutrients' enrichment is considered as one of the most significant reasons of EU MS water types' degradation. Eutrophication through nutrients' enrichment is also endangering the essential ecosystem services, including the supply of drinking water, the safe recreation activities, the protection of ecologically significant habitats, as well as the flora and fauna physiology of living [43].

Subsequently, the EU WFD supports the integrated assessment of ecological status through (a) biological quality elements of phytoplankton, benthic flora, benthic invertebrate and fish fauna, and (b) supporting the hydromorphology and the physicochemical parameters, including that of nutrients' profiling. Based on the EU WFD principles, the "good ecological status" implies that nutrient concentrations must "*not exceed the levels established so as to ensure the functioning of the ecosystem and the achievement of values specified (for good status) for the biological quality elements*" (Annex V, 1.2).

While the EU WFD is not setting specific nutrient concentration targets, it is requiring EU MS to set those nutrient criteria to ensure and support good ecological status. While there is literature reported a thriving research production devoted to the development of the assessment methods using Biological Quality Elements (BQEs) [44], sparse literature interest has been focused on, just only recently setting, nutrient criteria [45]. While a commonly known viewpoint of "good status" is appreciated through the requirement of inter-calibrating the biological elements, this process was not required for these elements involved.

3.5.2 Mitigating Need of Eutrophication Nuisance to Satisfy the Criteria of Water Framework Directive

The implementation of the EU WFD proved a credible tool that attracted the contemporary research interest in maintaining an acceptable water quality. However, during the running decades of the second half of the twentieth century there are, once oligotrophic, waters that have been gradually shifted to turbid- and phytoplankton-dominated systems [46]. The commonly, cyanobacteria-dominated, phytoplankton blooms are attributed to the crops' fertilization with nutrients, followed by nutrients' input to local surface water and groundwater. A plethora of intervention (post-treatment liking) management techniques, e.g., flushing, mixing, algicide-applications and biomanipulation, can effectively mitigate the negative effects of eutrophication in the short-run, but only the (proactive liking) treatment of a substantial reduction to the total nutrient loading could effectively accomplish a long term restoration from waters' eutrophication [10, 19].

While the phytoplankton growth is attributed to nutrients' inflows, lakes' recovery is predominately related to P-reduction [11, 14], indicating and proving the Golterman's statement, being said at the year 1975, that: "*It is not important whether phosphate is currently the limiting factor or not, or even that it has ever been so; it is the only essential element that can easily be made to limit algal growth*". P-reduction implies the loads' reduction of both external and internal P-inflows [46]. Besides, in lakes created by excavate-created lakes, the internal P-loading is often the main driver to sustain phytoplankton growth and to retard recovery for a substantial period after the external P-loading reduction [47]. While at P-enriched sediments, the P removal is a common method of internal P-loading reduction, it shows the constraints of (a) not being always effective [48] and (b) it is considered expensive in terms of recovery as well as for dewatering and long term disposal [48].

This phenomenon explains the need of adopting alternative methods in order to control the sediment-based P-release (being released as phosphate) and internal P-recycling in lakes. Consequently, there is a growing interest in P-sorbing materials [49, 50]. Among them, it has been proposed a novel method applied to a former sand excavation site, Lake Rauwbraken, The Netherlands, that has substantially burdened by intensive cyanobacterial blooms since 2000, before this official bathing site to undergo restoration [51]. The system analysis of this lake disclosed that internal P-loading was almost ten times higher than external P-input. Besides, regarding the high portion of diffuse inputs, being mostly referred to groundwater, it developed a joint capping treatment of phytoplankton flocculation/precipitation and sediment. It is noteworthy that the cyanobacteria biomass can be effectively removed from the water column while applying a low dose of polyaluminiumchloride (PAC) as flocculant with increased density, being fed by two tonnes of a Lanthanum Modified Bentonite (LMB) [51].

LMB was developed by CSIRO Australia and it is characterized by high binding capacity for P [52]. In particular, the settled matter and the sediment can be covered by an active barrier of the LMB designed to capture P-release from decaying settled

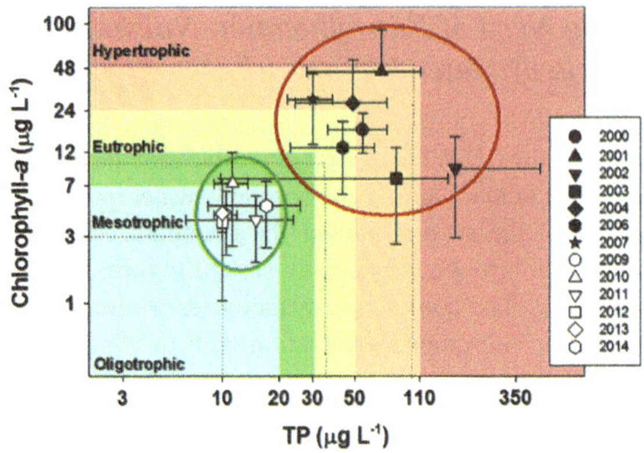

Fig. 3.10 The profile of EU WFD classification based on mean summer concentrations of total P (TP, mg L^{-1}) and chlorophyll-a (mg L^{-1}). The coloured palette indicates the EU WFD status: blue high, green good, yellow moderate, orange poor and red bad. Filled-legend symbols: Before Flock and Lock treatment, for the period 2000–2007. Open-legend symbols: After Flock and Lock treatment, for the period 2009–2014. *Source* Adapted from [46] with the permission from Elsevier

matter, sediment and incoming groundwater [51]. Such an LMB high P-removal capacity from the water column during settling is able to intercept P-release from sediments. It is noteworthy that the strong P-binding capacity of LMB is as follows:

- It is not affected by anoxia [53].
- It is making LMB to support a proven efficacy as a promising material to application in lakes where anoxic events are annually occurred near the sediment, as that of the stratified Lake De Kuil, where a modified form of LMB was applied [46].

An interesting profile of EU WFD classification based on mean summer concentrations of total P (TP, mg L^{-1}) and chlorophyll-a (mg L^{-1}) is represented in Fig. 3.10, where the recreational Lake De Kuil was valued under the EU WFD applicability [46]. It was also reported that at the case of Lake Rauwbraken, the LMB was applied at the lake's surface, and at the case of Lake De Kuil, most of the LMB was applied at deep injection. In such a way, the placement of LMB can improve the lake's bottom positioning and it can simultaneously prevent turbid waters due to surface application.

The aforesaid treatment is targeted to the sediment P-release and to the concentrations of TP and chlorophyll-a. After applying the Flock and Lock treatment, it was observed that the trophic level of the Lake De Kuil, being based on TP and chlorophyll-a, was shifted from the eutrophic to the mesotrophic status (Fig. 3.10). Moreover, it reported a shift from the "bad-poor moderate" to the "high" status, based on EU WFD classification of concentrations for summer TP and chlorophyll-a (Fig. 3.10). The reported shift has ensured a six summers' stability that indicated the positive long-term effects of the treatment adopted. Additionally, in monitoring at the long-run, it observed a year-to-year fluctuation in (a) TP concentration, (b) chlorophyll-a concentration, and (c) the Secchi depth.

3.6 Conclusions

In this chapter it deployed a systematic approach to, first, determine the sources and the causes of eutrophication-burdened water sources and, second, provide proactive and post-treatment restoration techniques that are certainly abiding by legal frameworks and regulatory Directives among EU MS. The analysis has been also enriched by the literature coverage of specific paradigms that govern water protection in European and international contexts. The critical issues of local and nations' responsibility of rebounding the healthy conditions of their water sources are also determined by the knowledge level of experience, the innovative diffusion of remediation techniques, as well as the confrontation to relevant legislative frameworks, such as the EU Water Framework Directive 2000/60/EC. The key-aspects of climatic conditions, temperature variations, anthropocentric activities, and the sharing responsibility of citizens and central governments to eutrophication reduction have been proven critical indicators of (a) reducing the internal sources of phosphorus from the sediment, (b) improving water quality, (c) controlling the water eutrophication levels in the long-run while simultaneously (d) achieving environmental conditions of sustainability through abiding to national and EU MS legislative frameworks.

References

1. Water Framework Directive (2000) Directive 2000/60/EC of the European Parliament and of the Council establishing a framework for Community action in the field of water policy. Off J Eur Parliam
2. Directive 2005/56/EC of the European Parliament and of the Council. In: Cross-border mergers in Europe, 2011
3. Resources, C. on E. and N. (2010) Scientific assessment of hypoxia in U.S. coastal waters. Interagency Working Group on Harmful Algal Blooms, Hypoxia, and Human Health of the Joint Subcommittee on Ocean Science and Technology
4. Voulvoulis N, Arpon KD, Giakoumis T (2017) The EU Water Framework Directive: from great expectations to problems with implementation. Sci Total Environ
5. European Environment Agency (2015) European briefings—climate change impacts and adaptation. In: SOER 2015—the European environment—state and outlook 2015
6. European Commission introduction to the new EU Water Framework Directive—Environment—European Commission. https://ec.europa.eu/environment/water/water-framework/info/intro_en.htm
7. European Commission (2019) European Union strategic approach to pharmaceuticals in the environment
8. European Environmental Agency (2018) European waters assessment of status and pressures 2018
9. Pietrzak D, Kania J, Malina G, Kmiecik E, Wątor K (2019) Pesticides from the EU First and second watch lists in the water environment. Clean Soil Air Water

10. Zamparas M, Kyriakopoulos GL, Drosos M, Kapsalis VC, Kalavrouziotis IK (2020) Novel composite materials for lake restoration: a new approach impacting on ecology and circular economy. Sustainability
11. Zamparas M, Drosos M, Kalavrouziotis IK (2019) Sorption of phosphate from innovative composite material focusing on physicochemical interactions. Desalin Water Treat
12. Peeters M, Eliantonio M, Hervé-Fournereau N (2020) Beyond the 2019 fitness check of the Water Framework Directive: designing the future of European water law. In: Research handbook on EU environmental law
13. Zal N, Whalley C, Christiansen T, Kristensen P, Néry F, Whalley C, Zal F, Christiansen T, Schmedtje U, Solheim AL et al (2018) European waters—assessment of status and pressures. European environment agency report 7/2018. ISBN 9789292139476
14. Zamparas M, Deligiannakis Y, Zacharias I (2013) Phosphate adsorption from natural waters and evaluation of sediment capping using modified clays. Desalin Water Treat
15. Solheim AL, Rekolainen S, Moe SJ, Carvalho L, Phillips G, Ptacnik R, Penning WE, Toth LG, O'Toole C, Schartau AKL et al (2008) Ecological threshold responses in European lakes and their applicability for the Water Framework Directive (WFD) implementation: synthesis of lakes results from the REBECCA project
16. Poikane S, Kelly MG, Salas Herrero F, Pitt JA, Jarvie HP, Claussen U, Leujak W, Lyche Solheim A, Teixeira H, Phillips G (2019) Nutrient criteria for surface waters under the European Water Framework Directive: current state-of-the-art, challenges and future outlook. Sci Total Environ
17. European Environment Agency (2006) Progress towards halting the loss of biodiversity by 2010. EEA report
18. Kagalou I, Papastergiadou E, Leonardos I (2008) Long term changes in the eutrophication process in a shallow Mediterranean lake ecosystem of W. Greece: response after the reduction of external load. J Environ Manag 87:497–506
19. Zamparas M, Zacharias I (2014) Restoration of eutrophic freshwater by managing internal nutrient loads. A review. Sci Total Environ 496:551–562
20. Gianni A, Zamparas M, Papadas IT, Kehayias G, Deligiannakis Y, Zacharias I (2013) Monitoring and modeling of metal concentration distributions in anoxic basins: Aitoliko Lagoon, Greece. Aquat Geochem 19:77–95
21. Wang C, He R, Wu Y, Lürling M, Cai H, Jiang HL, Liu X (2017) Bioavailable phosphorus (P) reduction is less than mobile P immobilization in lake sediment for eutrophication control by inactivating agents. Water Res
22. Lürling M, Mucci M (2020) Mitigating eutrophication nuisance: in-lake measures are becoming inevitable in eutrophic waters in the Netherlands. Hydrobiologia
23. OECD (2014) Water governance in the Netherlands
24. European Commission (2018) Proposal for a Directive on the quality of water intended for human consumption. https://ec.europa.eu/environment/water/water-drink/review_en.html
25. Ansari AA, Gill SS (2014) Eutrophication: causes, consequences and control, vol 2. ISBN 9789400778146
26. Andersen JH, Conley DJ (2009) Eutrophication in coastal marine ecosystems: towards better understanding and management strategies. Hydrobiologia
27. Van Der Brugge R, Rotmans J (2007) Towards transition management of European water resources. In: Proceedings of the integrated assessment of water resources and global change: a North-South analysis
28. Volk M, Liersch S, Schmidt G (2009) Towards the implementation of the European Water Framework Directive? Land Use Policy
29. Martin-Ortega J (2012) Economic prescriptions and policy applications in the implementation of the European Water Framework Directive. Environ Sci Policy
30. Scottish Environmental Protection; Water (2011) The water environment (controlled activities) (Scotland) regulations 2011. Regulations
31. Giakoumis T, Voulvoulis N (2018) The transition of EU water policy towards the Water Framework Directive's integrated river basin management paradigm. Environ Manag
32. European Commission (2006) EU Groundwater Directive 2006/118/EC

33. European Union (2007) Directive 2007/60/EC of the European Council and European Parliament of 23 October 2007 on the assessment and management of flood risks
34. European Commission (2012) Proposal for a Directive of the European Parliament and of the Council amending Directives 2000/60/EC and 2008/105/EC as regards priority substances in the field of water policy. ISBN 0060200901
35. Petersen JD, Rask N, Madsen HB, Jørgensen OT, Petersen SE, Nielsen SVK, Pedersen CB, Jensen MH (2009) Odense Pilot River Basin: implementation of the EU Water Framework Directive in a shallow eutrophic estuary (Odense Fjord, Denmark) and its upstream catchment. In: Eutrophication in coastal ecosystems
36. Verbeeck B, Leroy P (2006) A target group approach in Flemish environmental policy: establishing new government-industry relations? In: Institutional dynamics in environmental governance. ISBN 140205078X
37. Vlachopoulou M, Coughlin D, Forrow D, Kirk S, Logan P, Voulvoulis N (2014) The potential of using the ecosystem approach for WFD implementation. Sci Total Environ
38. Steyaert P, Ollivier G (2007) The European Water Framework Directive: how ecological assumptions frame technical and social change. Ecol Soc
39. Sabel CF, Zeitlin J (2012) Experimentalism in the EU: common ground and persistent differences. Regul Gov
40. European Commission (2000) Priority substances under the Water Framework Directive. Directive 2008/105/EC
41. van Rijswick HFMW, Backes CW (2015) Ground breaking landmark case on environmental quality standards? J Eur Environ Plan Law
42. Boscheck R (2006) The EU Water Framework Directive: meeting the global call for regulatory guidance? Intereconomics
43. Culhane FE, Briers RA, Tett P, Fernandes TF (2014) Structural and functional indices show similar performance in marine ecosystem quality assessment. Ecol Indic
44. Caroni R, van de Bund W, Clarke RT, Johnson RK (2013) Combination of multiple biological quality elements into waterbody assessment of surface waters. Hydrobiologia
45. Herrero FS, Teixeira H, Poikane S (2019) A novel approach for deriving nutrient criteria to support good ecological status: application to coastal and transitional waters and indications for use. Front Mar Sci
46. Waajen G, van Oosterhout F, Douglas G, Lürling M (2016) Management of eutrophication in Lake De Kuil (The Netherlands) using combined flocculant—lanthanum modified bentonite treatment. Water Res
47. John Anderson N, Jeppesen E, Søndergaard M (2005) Ecological effects of reduced nutrient loading (oligotrophication) on lakes: an introduction. Freshw Biol
48. Reddy KR, Fisher MM, Wang Y, White JR, James RT (2007) Potential effects of sediment dredging on internal phosphorus loading in a shallow, subtropical lake. Lake Reserv Manag
49. Hickey CW, Gibbs MM (2009) Lake sediment phosphorus release management—decision support and risk assessment framework. N Z J Mar Freshw Res 43:819–856
50. Zamparas M, Drosos M, Deligiannakis Y, Zacharias I (2015) Eutrophication control using a novel bentonite humic-acid composite material Bephos^{TM}. J Environ Chem Eng 3:3030–3036
51. Lürling M, Van Oosterhout F (2013) Controlling eutrophication by combined bloom precipitation and sediment phosphorus inactivation. Water Res
52. Zamparas M, Gianni A, Stathi P, Deligiannakis Y, Zacharias I (2012) Removal of phosphate from natural waters using innovative modified bentonites. Appl Clay Sci
53. Hickey CW, Gibbs MM (2009) Lake sediment phosphorus release management-decision support and risk assessment framework. N Z J Mar Freshw Res

Chapter 4
Management Strategies for Lake Restoration

Miltiadis G. Zamparas

Abstract Nowadays in many lakes, even if adequate external P load reductions have been achieved, internal P load reductions to limit nutrients inflows are required for positive outcomes within reasonable time scales. In accelerating the recovery of lakes' health through external nutrient reduction, the in-lake methods are commonly affected. Such methods include physical, chemical, and biological decrease in internal loading in many water bodies, which have shown relative success. In this chapter, the research objectives are targeted to (a) discuss the role of internal nutrient loading in the eutrophication process, (b) describe the restoration mechanisms and methods (focusing on physical methods) while evaluating their relative successes and failures in controlling eutrophication, and (c) review each restoration method to better understand scientific knowledge on their ecological footprint, thus recommending those restoration methods matching to each lake-case.

M. G. Zamparas (✉)
Laboratory of Sustainable Waste Technology Management, School of Science and Technology, Hellenic Open University, Parodos Aristotelous 18, 26335 Patras, Greece
e-mail: mzamparas@upatras.gr

M. G. Zamparas and G. L. Kyriakopoulos (eds.), *Chemical Lake Restoration*,
https://doi.org/10.1007/978-3-030-76380-0_4

Graphical Abstract

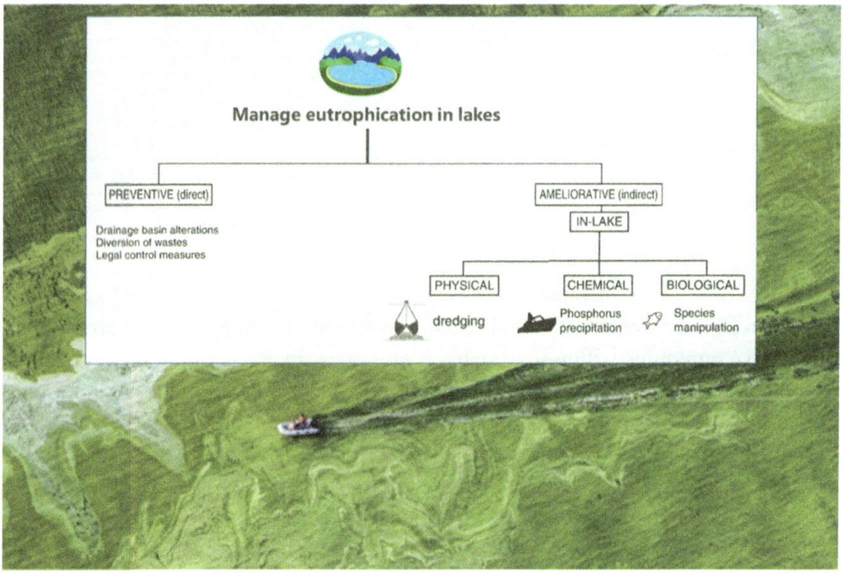

Graphical Abstract: Management panorama of eutrophication in lakes

Keywords Eutrophication control · Dredging · Biomanipulation · Hypolimnetic withdrawal · Aeration · P-inactivation agents

4.1 Introduction

4.1.1 The Role of External and Internal Nutrient Loads in the Eutrophication Process

One of the most crucial components in phytoplankton biomass is nutrient loads, which refer to the total amount of nutrients (P&N) inflowing a water body over a given period of time [1]. Both internal and external sources of nutrients are available in the water column. Natural contributions from rain and runoff, as well as human-caused inputs from diffuse sources like fertilizer inputs and point sources like sewage treatment plants, are all external sources of nutrients [2, 3]. Internal loading is a term that refers to the recycling of nutrients from sediments, which occurs primarily through redox reactions at the sediment–water interface and the decomposition of settling organic matter [4]. Internal recycling is predominant in lakes that undergo stratification and hypolimnetic oxygen depletion (Fig. 4.1). Firstly, remediation of lakes from eutrophication should start with the decrease in nutrient inputs from external sources [5].

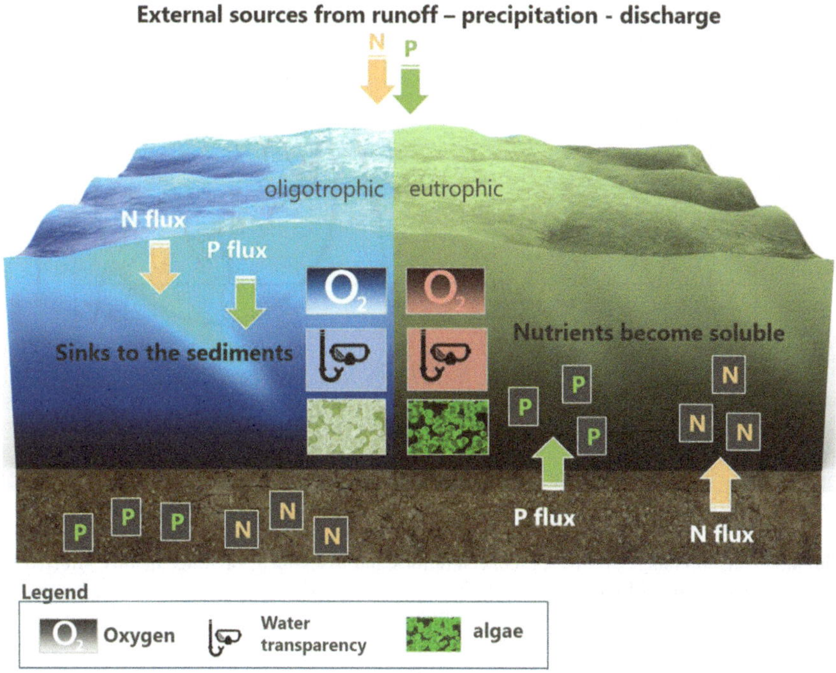

Fig. 4.1 P and N elemental recycling in the sediment–water interface

Otherwise, lakes' eutrophication is deemed to be short-sighted and constrained by controlling nutrients and organic matter inflows of lakes caused by these external sources [6, 7]. In parallel, increased quantities of P-flux from bottom sediments are stated as it is considered the main preventing mechanism of fast lake recovery after an external loading reduction [5].

Based on the above findings it is noteworthy the pronounced role of sediment in good water quality of lakes, balancing out the competitive situation of both nutrient source and sinking sedimentation rate. In this respect, Lee et al. [8] stressed the dependence of water amounts in phosphorus and nitrogen with internal loads from the sediment (Fig. 4.2). In particular, these authors developed a three-dimensional hydrodynamic and transport model on a benthic chamber to inspect the effects of sediment flux on the reservoir water quality at Euiam Lake, South Korea. This study verified the role of the sediment as a source or sink of nutrients, denoted that a variation of phosphorus flux rates can significantly extend the eutrophication period, and concluded the affective role among sediment management of nutrients and contaminants on water quality, being proven an effective operation tool especially in treating serious algal bloom throughout spring and fall (Fig. 4.2) [8].

Plentiful studies have been focused on examining the role of the benthic fluxes in the light of environmental changes. However, only a few studies have been examined the regulative role of internal benthic loads in the elemental balance levels at the whole

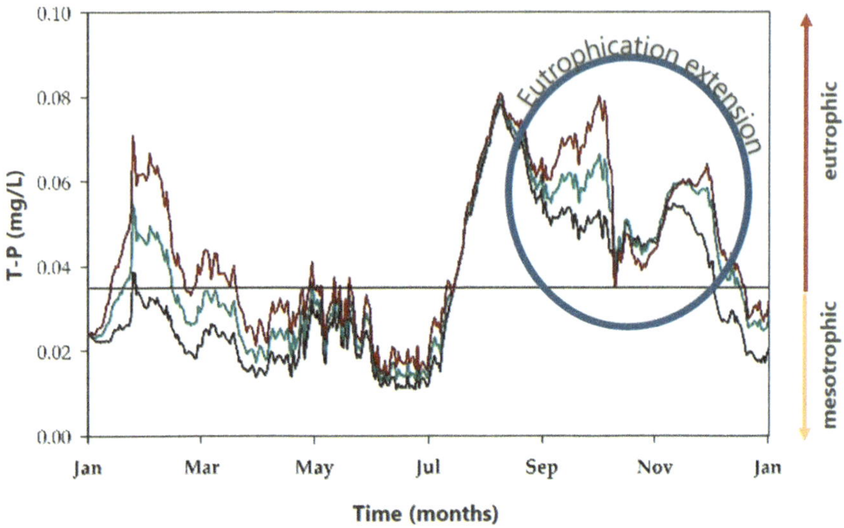

Fig. 4.2 Annual changes in the TP concentrations and assessment of water quality affected by sediment flux. *Source* Modified from [8]

lakes. Therefore, the elemental recirculation from sediment to the water column can be attributed to a wide range of water body characteristics, such as morphological (mean depth) [8–10], physical (including data on meteorology, advection flow, velocity, and thermal stratification) [11], chemical (mainly data on dissolved oxygen and redox potential) [12] and biological (bioturbation) [13].

In a similar study, a better understanding of the mobility and biogeochemistry of P under fluctuating redox conditions was studied by Parsons et al. [13] who denoted the high affinity of P recycling with redox regime in superficial sediment. Specifically, a sample of hypereutrophic freshwater marsh sediment was exposed to alternating seven-day periods of purging with air and nitrogen gas (N_2) in a bioreactor for 74 days. It showed the association between the magnitude and the timing of internal P loading during anoxia periods with both P redistribution in sediments and bottom-water nitrate concentrations [13].

Burger et al. [11] examined the inflows of internal and external nutrient fluxes and phytoplankton biomass in Lake Rotorua, New Zealand, through the DYRESM-CAEDYM vertically resolved water quality model. In this model, the relative theoretical water column temperature, soluble reactive phosphorus (SRP), and ammonium (NH_4) concentrations were proven closely matching to in-field measurements. The parameters of investigation were that of timing and period of stratification events as well as subsequent changes in nutrient concentrations of bottom water. Among the variable setting were data on surface water concentrations of chlorophyll-a, revealing that reductions in internal nutrient loads can be much more reduce-effective cyanobacterial biomass than similar reductions in catchment fluxes, as coinciding large sediment nutrient release events with high cyanobacterial biomass. It can be signified that substantial cyanobacterial biomass reduction in the studied lake could be achieved by a considerable and prolonged reduction in external loads, thus reducing internal loads in Lake Rotorua [11].

A longitudinal study among lakes in Europe and North America disclosed that sediment P release typically exists for more than 10–15 years, even more than 20 years, after the loading reduction [14]. The determinant role of sediments in lakes' eutrophication was notable in alignment with the adsorption and desorption capacity of a large number of toxic elements and compounds. Furthermore, physicochemical characteristics of the sediments and oxidation–reduction conditions between sediment and overlying water are linked to lake sediments' capability to retain P through controlling sediments' deposition, which must be carried out in order to achieve enhanced eutrophic remediation and lakes' recovery [15].

4.1.2 External Nutrient Loads Management

Reducing external nutrient loads is fundamental to sustainably restoring lakes degraded by eutrophication; a failure to enact sufficient reductions to external nutrient loads is a major cause of poor lake restoration outcomes [16]. The magnitudes of external load reductions required to achieve lake restoration goals are often underestimated, and managers should be wary of setting over-optimistic targets. Carpenter et al. [17] examined management policies for lakes subject to varying degrees of eutrophication. They grouped lakes in terms of responses to external load reductions, classifying lakes as either "*recoverable*" (recovery is immediate and proportional to phosphorus load reductions), "*hysteretic*" (very high reduction in phosphorus loads are required for a period), and "*irrecoverable*" (external phosphorus load reductions alone are insufficient to restore a lake) [17].

Shallow lakes were identified as most likely to correspond to the latter two categories. Policy analysis showed that the effects of stochasticity (e.g., in weather), lags, and uncertainty in limnological models mean that to maximize economic benefits of lake restoration, nutrient load reduction targets should be more ambitious than those indicated by empirical limnological models (e.g., TP—chl-a regressions) alone. There is a growing recognition that the adoption of Best Management Practices (BMPs) alone is insufficient to restore eutrophic lakes in catchments with high external nutrient loads from diffuse agricultural sources.

Osgood [18] reviewed the efficacy of BMPs to reduce external phosphorus loads to lakes in the USA, where BMPs are defined as "management techniques or landscape-modifying actions" such as detention basins, swales, constructed wetlands, filter strips, or porous pavements [16]. The author noted that phosphorus load reductions of >80% are required to restore most eutrophic lakes, yet the maximum possible phosphorus load reduction due to BMPs was ~50%, with reductions of <25% typically recorded in practice. The author concluded that the use of BMPs alone is commonly proven as an inadequate strategy at reversing eutrophication and advocated for restoration plans to also focus on internal loads by including in-lake engineering or chemical treatments, as these provide faster improvements, typically at a reduced cost [18]. This author estimated a long period of 80 years needed for Shagawa Lake in order a 90% recovery to be achieved at the P equilibrium concentration (termed as mean annual volume-weighted and whole-lake TP concentration) of 12 μg $L^{-1,}$ following an almost 80% external P loading reduction. Moreover, Osgood [18] was forecasting a 250 year time lag period for the recovery of Yahara (WI) lakes.

Even in lake-cases of successful control of external P load reductions, a reasonable timeline of controlling internal P load reductions is also required. Such controlling agents of internal P load reductions are that of P-inactivation agents, e.g., aluminum sulfate, while new precipitants of LMB, Z2G1 have proven evaluation and operation for long-term internal P loading [19–22]. In the relevant literature, similar researches on in-lake management at internal P loading mitigation were that of hypolimnetic withdrawal [23], biomanipulation [24], and dredging [25], which have a long track record of effectiveness.

4.2 Lake Sediment Release Management—Overview of Restoration Procedure

4.2.1 Dredging

Dredging involves removing lakebed sediments and therefore has the advantage of directly removing nutrients from a lake that have accumulated due to the legacy of high external loading [25–28]. In shallow lakes, dredging is a primary method to physically control internal loads [29]. While dredging can moderate internal loadings, it can also effect on changing long-run physicochemical conditions at the sediment–water interface [28], including changes at exposing subsurface sediments to the surface, reduction in oxygen depletion, aeration state (Eh) increase, pH and sediment porosity lowering [28].

Ammonia regeneration (termed ammonification) from organic-N in sediments can be weakened after dredging (Fig. 4.3), due to relative oxic conditions prevailing in sediments from dredged groups. Besides, due to the fact of low bioavailability of organic nitrogen to the growth of algal and aquatic plants, dredging can be a countermeasure for lakes to mitigate ammonia release while mitigating the regeneration of ammonia and reducing the depletion of oxygen (Fig. 4.3).

Dredging can substantially reduce sediment nutrient releases; for example, a six-month sediment incubation experiment using samples from hypertrophic Taihu Lake (mean depth = 2 m) showed that maximum phosphorus release was reduced significantly [30, 31]. The fluxes of soluble reactive phosphorus (SRP) from the

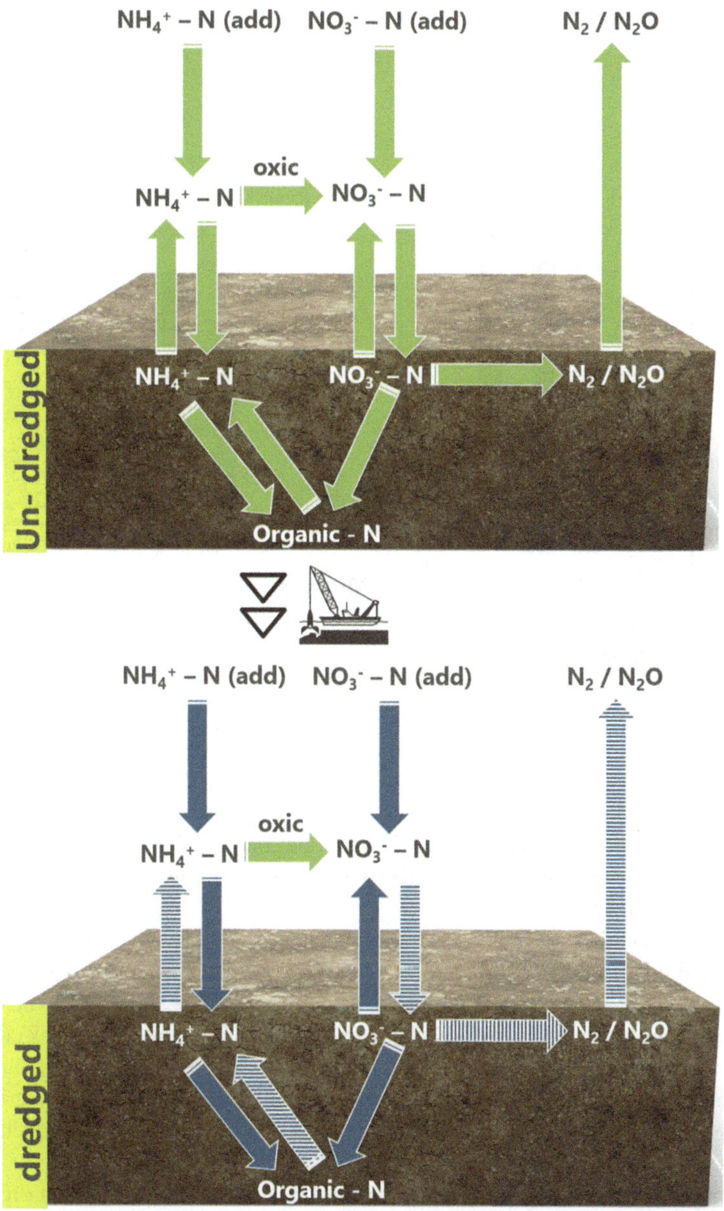

Fig. 4.3 Nitrogen equilibrium in un-dredged and dredged sediment-water environments. Dashed arrow in dredged sediments means this procedure was relatively weak related to un-dredged sediments. *Source* Modified from [28]

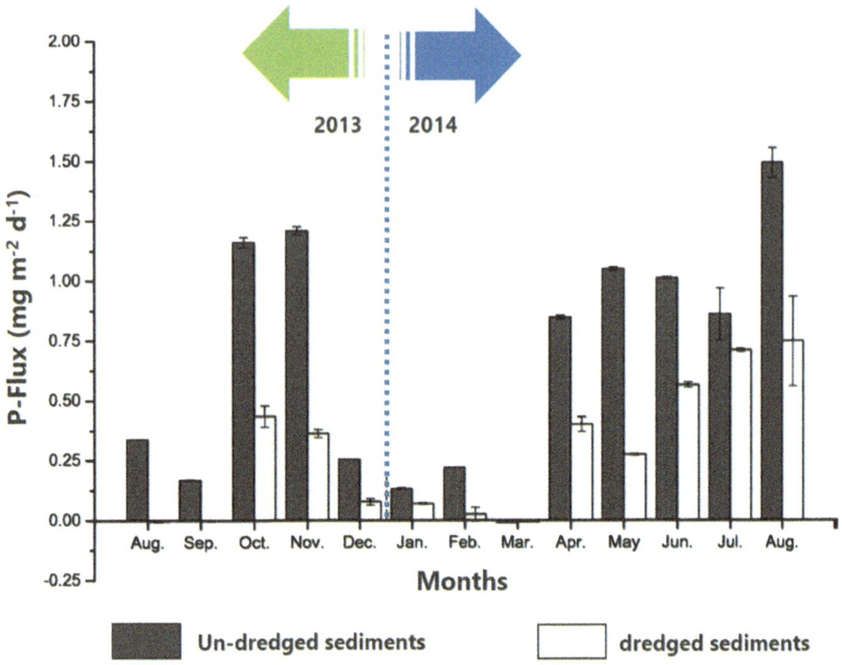

Fig. 4.4 P-fluxes across the sediment–water interface from August 2013 to August 2014. *Source* Modified from [31]

post-dredged sediments were remarkably reduced by 58%, contrasted with the non-dredged sediments. Furthermore, the P-fractionation revealed that the contents of the different P-forms and the TP of the post-dredged sediments were significantly decreased [31] (Fig. 4.4).

Moreover, another study at Taihu Lake confirmed that reduced phosphorus release due to dredging results in reduced soluble reactive phosphorus (SRP) concentrations in the overlying water; field sampling there showed that peak surface water SRP concentrations at dredged sites (~50–55 μg/L) were slightly lower than peak concentrations at an un-dredged site (62 μg/L) [32]. This reduction in the water column of SRP concentrations was markedly lower than indicated by the laboratory experiment, although there was presumably some mixing between dredged and un-dredged areas [32].

Zhong et al. [33] recently published a report that evaluated the effect of dredging season on sediment properties and nutrient fluxes at two contaminated sites in Meiliang Bay and Taihu Lake (China). Dredging the first 30 cm of sediment will reduce organic matter, total nitrogen, and total phosphorus, according to the experimental design, in various seasons. In both the Inner and Outer Bays, dredging had a major restricting impact on P release, and soluble reactive phosphorus (SRP) fluxes from the dredged cores were typically lower (p 0.05) than that from the un-dredged cores. NH_4^+-N fluxes from the dredged cores were considerably higher ($p < 0.05$) than that from the un-dredged cores in all seasons for both areas, indicating short-term risk of NH_4^+-N release after dredging, and this risk is greatest in seasons at higher temperatures as that at the Inner Bay (Fig. 4.5). Therefore dredging could effectively address internal loading in Taihu Lake, and that the seasons with low temperature are proper for implementing dredging projects [33].

Removing nutrient-rich sediments is the main mechanism by which dredging reduces internal loads resulting in stark ecological improvements in shallow lakes [34]. Increasing the lake depth will also reduce the rate of wind-driven resuspension. This was demonstrated in a modeling study of a eutrophic peat lake in the Netherlands (mean depth = 1.77 m), which showed that increasing the depth to 12 m in localized areas that cover 10% of the lakebed would reduce suspended particulate organic sediment by 25% and improve water clarity [35]. One further advantage of dredging is the potential to reduce the prevalence of cyanobacteria by the direct removal of akinetes (resting cells) from the sediments [36].

Today, dredging has limited use. This may reflect the high costs, meaning that it is most feasible for small lakes and those that are in high-use urban areas. Moreover, sediments suspended during the dredging process may also remain in the water column for extended periods (~days–weeks). This can temporarily reduce clarity and mobilize contaminants present in the sediments such as organochlorine pesticides or polychlorinated biphenyl (PCB) compounds. The presence of contaminants can pose difficulties for the disposal of dredged sediments. The 'dredge-skim' method can potentially alleviate the need to dispose of contaminated sediments; this involves pushing the upper layer of sediments to one side of the lake during dredging and only removing the deeper, potentially less-contaminated material [37]. In cases where sediments are uncontaminated, the removal of nutrient-rich sediments can provide a resource to amend agricultural soils. As with other methods to control internal loads, dredging is best undertaken in conjunction with external load reduction as failure to reduce external nutrient loads will provide only temporary results [37].

Hickey and Gibbs [38] stated that the dredging method can directly remove sediment thus reducing the internal flux, but basic knowledge on sediment composition is needed to guarantee such a sufficient removal of depth forming the high-nutrient sediment. Dredging is more suitable for small lakes but critical issues and practical consideration that have, yet, to be addressed are that of how much sediment has to be removed (mainly from the depth and from the nearby-to-lake site), where contaminated sediment should be disposed after in-field extraction/removal, as well as the feasibility cost at implementing such a considerable amount of sediment removal. Summarily the main disadvantages are (1) the impossibly elevated cost of sediments disposal for local rural use at chemical contaminant-rich sediments (e.g., urban runoff, geothermal contaminants); (2) the large and specialized equipment at large-scale treatments; and (3) the significant disturbance and contaminant effluents into lake water during dredging procedure (Fig. 4.6). On the other hand, this approach supports advantageous features, such as that of (1) permanent removal of contaminant-rich sediments at small-areas, being adversely affecting the lake ecosystem; and (2) cost recovery may be possible for non-geothermally influenced lakes, when sediment is the feedstock fertilizer source for agriculture [38].

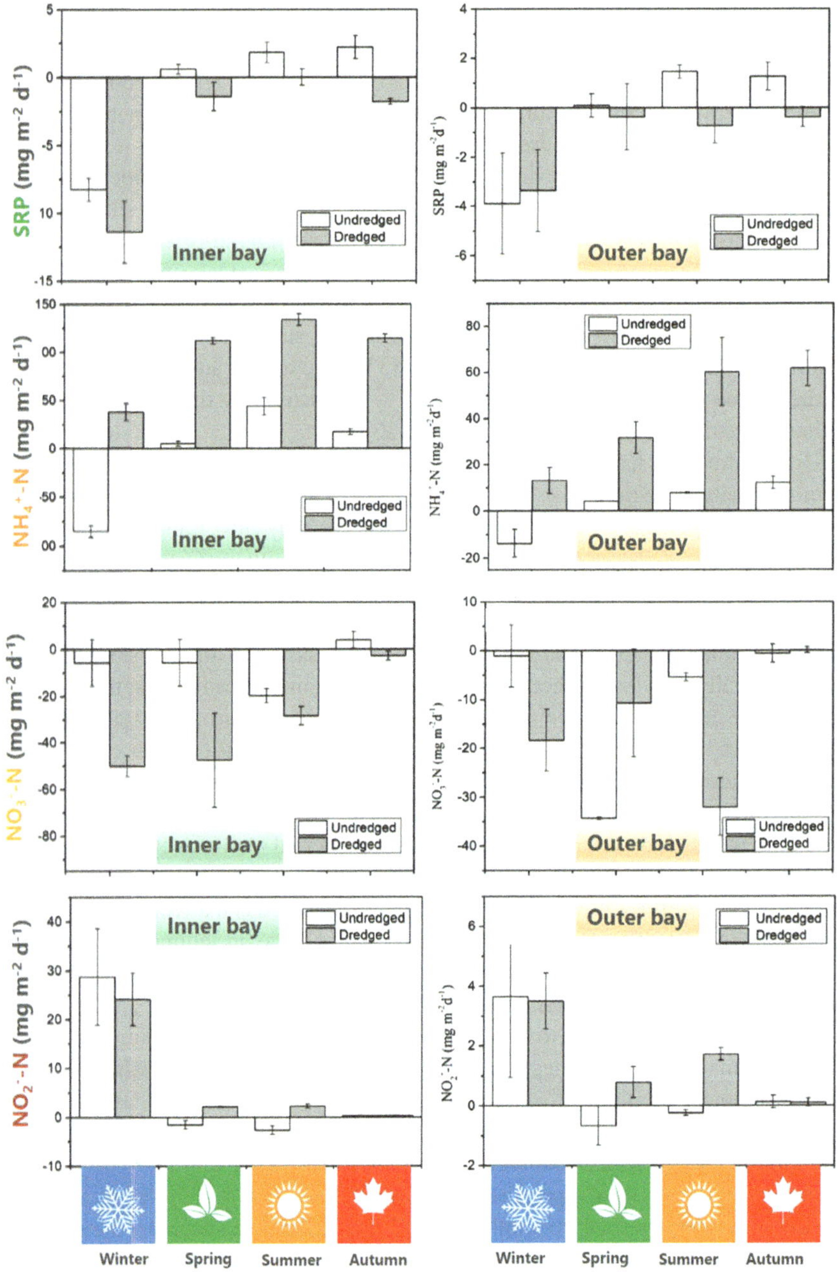

Fig. 4.5 SRP and inorganic nitrogen fluxes across the sediment–water interface in un-dredged and dredged sediments of the Inner Bay and Outer Bay. *Source* Modified from [33]

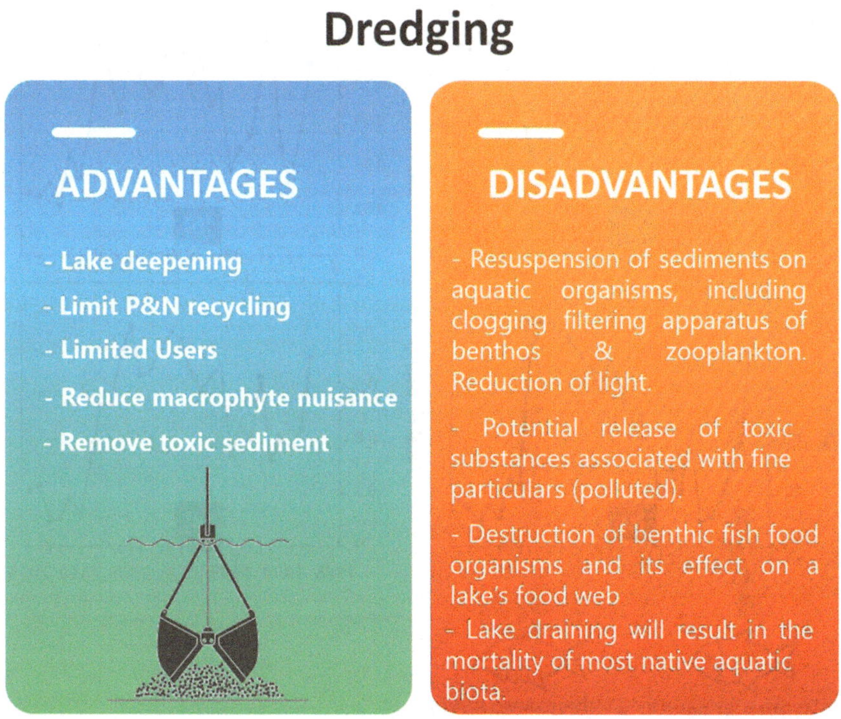

Fig. 4.6 Advantages and disadvantages of dredging in shallow lakes. *Source* Author's own study

4.2.2 Biomanipulation

Among the methods applied for the improvement of lake water quality is that of biomanipulation by fish removal. Søndergaard et al. [39] studied the shallow and 16 ha large Lake Væng (Denmark) by utilizing chemical and biological data of 30 years. This water source was biomanipulated twice with a 20-year interval by removing roach (*Rutilus rutilus*) and bream (*Abramis brama*). Following these two types of biomanipulation, the turbidity- and phytoplankton-dominated Lake Væng shifted to clear- and water macrophyte-dominated conditions. Consequently, the reduction from 60–80 to 10–30 $\mu g\ L^{-1}$ for chlorophyll was reported (Fig. 4.7), and an increase in submerged macrophytes from <0.1 to 70–80% by, mainly, *Elodea canadensis* coverage was reported. Besides, two approximated 0.12–0.07 and 1.0–0.4 $mg\ L^{-1}$ reductions in mean summer total phosphorus and total nitrogen were reported, respectively (Fig. 4.7). From a seasonal point of view, summer maximum concentrations of phosphorus and chlorophyll during turbid conditions have been changed to winter maximum under clear conditions. The relatively high phosphorus loading via the groundwater and the accumulation of a mobile P pool in the sediment are making uncertain the future of Lake Væng since it is likely that the lake eventually transited to turbid conditions. A promising strategic management of repeated fish

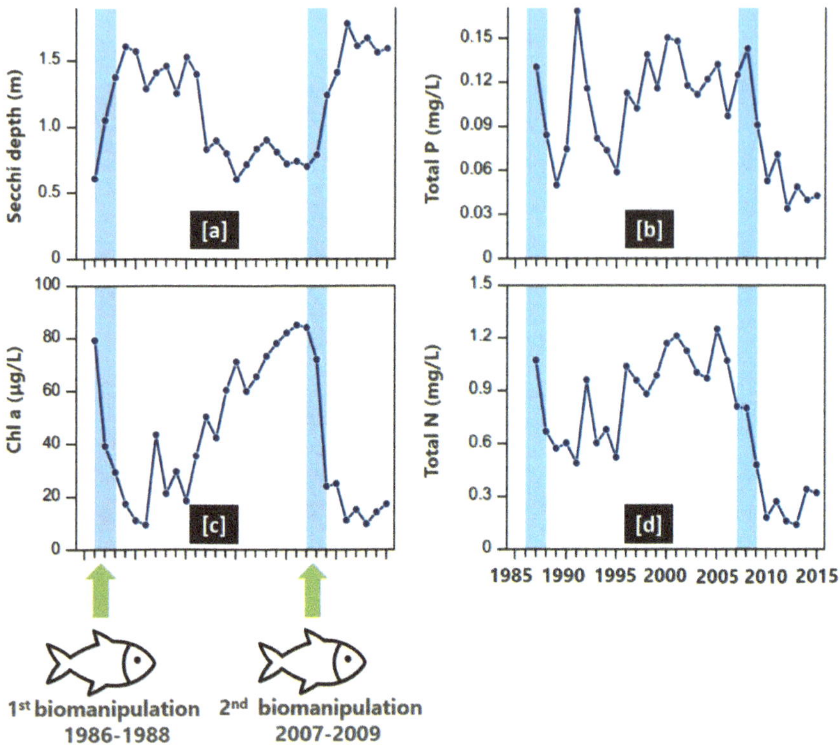

Fig. 4.7 Mean summer Secchi depth, Chl-a, and concentrations of nutrients (TN and TP) from 1986 to 2015. The two periods with fish removal are indicated with blue color. *Source* Modified from [39]

removals might apply to shallow lakes, having a relatively high loading of external nutrients.

The low cost and decreased algal concentrations are two advantages of this approach. This approach is non-invasive (as opposed to dredging) and may be called "natural." The key drawback to this approach is that even though it is effective, in the sense that zooplankton populations increase and algae populations are extensively grazed, algae populations would not be drastically decreased. Even if algae levels are lowered, surface water phosphorus concentrations cannot reach the Bassett Creek Watershed Management Commission's (BCWMC) water quality targets or State water quality requirements (Fig. 4.8).

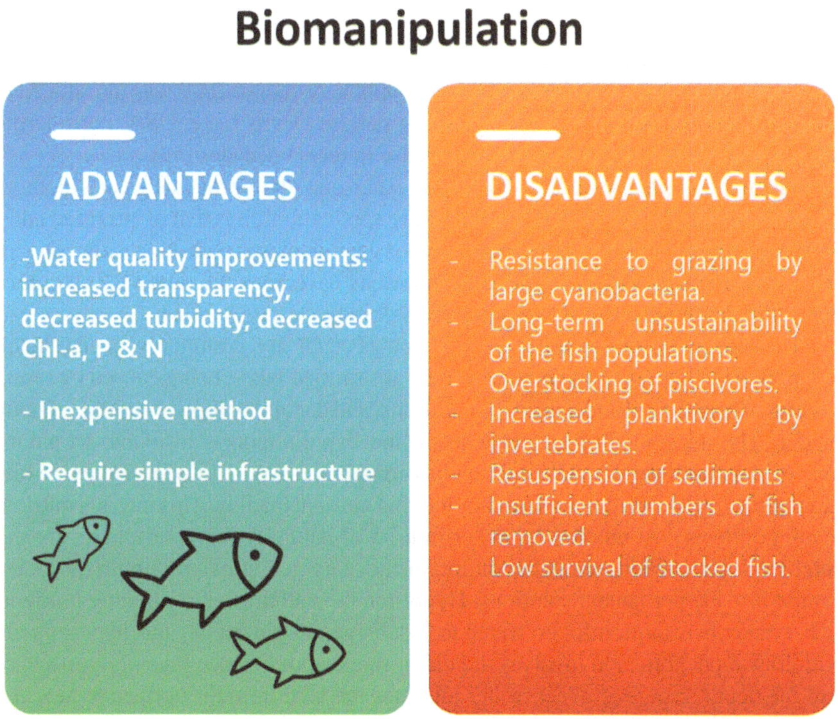

Fig. 4.8 Advantages and disadvantages of biomanipulation in shallow lakes. *Source* Author's own study

4.2.3 Hypolimnetic Withdrawal

The microbial decomposition of organic substances eliminates dissolved oxygen in the deep water can develop an anoxic hypolimnion at thermally stratified lakes [40]. At hypolimnia, the underlying sediments are commonly considered as sources of phosphorus (P), since the accumulation of legacy P is causing enrichment at easily available P sediment. However, it is problematic when reduction input of external P is made without satisfactory improvement of the trophic state and the water quality. Therefore, the influential role of internal P loading on water quality can be decreased by the lake restoration treatment.

The hypolimnetic withdrawal installation has been successfully documented in various regions worldwide including the USA, Canada, Germany, Finland, Austria, Poland and Australia [23, 41, 42]. At almost all these studied lakes, a maximum hypolimnetic TP concentration, as well as epilimnetic TP concentration, has been decreased. The change in epilimnetic TP concentrations was time-depended since the longer the withdrawal was operating, the greater was the change in epilimnetic TP concentrations. Hypolimnetic water dispositions, being enriched with nutrients, have been reported in the mid-twentieth century and particularly measured while using a hypolimnetic withdrawal siphon. This method has been designed to operate during September up to early October (where the highest hypolimnetic P concentrations have been measured), while a modification siphon system in 2009 was introduced to support flood prevention in the surrounding park in Devils Lake, USA, considering the regionally dramatic precipitation increase in recent decades [43].

In general, this technique is suitable for water-quality improvements at a plethora of applications [23, 29, 41, 43]. Among these applications, it is that of stratified lakes with hypolimnia enriched in nutrients and oxygen-depleting substances. In lakes of HW, hypolimnetic maximum TP concentrations have been measured 2.2–90 times that of the epilimnion during summer and fall [44].

Phosphorus, ammonia, and ferrous iron (Fe(II)) are commonly accumulated compounds in hypolimnetic bottom waters at stratified lakes, being caused by redox cycling of iron (Fe), P under anoxic conditions, and decomposition of settling organic matter. The accumulation of nutrients in the hypolimnion is available to phytoplankton through periodic lake mixing or vertical migration of phytoplankton, while phytoplankton is considered as a major internal nutrient loading. This internal nutrient loading reflects past nutrient inputs or inflows of naturally high nutrient levels and catchment soils, while remediation approaches targeted at external nutrient loads, cannot provide permanent solutions. Hypolimnetic withdrawal is allowing nutrient-rich rather than nutrient-poor water to discharge. In particular, the hypolimnetic withdrawal phenomenon involves the direct removal of P-laden lake bottom waters and the depth change at which water migrates the lake from the surface to near the maximum depth (Fig. 4.9).

This is usually accomplished by constructing an outlet weir (Olszewski Pipe) and replacing outflowing surface water with hypolimnetic water (Fig. 4.9). As a result, the hypolimnion retention period is reduced, the likelihood of anaerobic conditions developing is reduced, and the availability of nutrients to the epilimnion is also

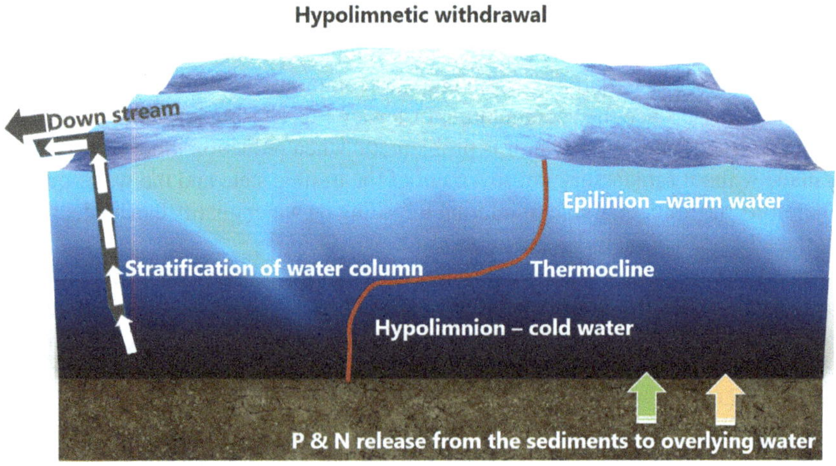

Fig. 4.9 Hypolimnetic withdrawal scheme

decreased through entrainment and diffusion [23, 43]. Installing a pipe along the lake's bottom from near the deepest point to the outlet is how the treatment is carried out. The outlet pipe, which acts as a siphon, is usually located below the lake level. This treatment is appropriate for stratified lakes and small reservoirs with anaerobic hypolimnia that restrict fish habitat while promoting the release of P, toxic metals, and hydrogen sulfide from sediments [43].

A successful treatment necessitates two important preconditions: First, the lake level must remain relatively constant and, second, thermal stability should not change [42, 43]. While epilimnetic water tends to be drawn downward, causing stratification weakening, destratification cannot be occurred, provided the slow removal rate of hypolimnetic water. Destratification is also increasing the transportation of hypolimnetic nutrients and anoxic water to the epilimnion; thus, it should be avoided. Changes' control due to destratification is related to the possible direction of inlet water to the metalimnion or hypolimnion. Preferentially removal of hypolimnetic water is causing a decrease in P internal loading through the residence time decrease in the hypolimnion and shortening of anoxia period, while the simultaneous increase in the anoxic boundary depth [42]. Hypolimnetic withdrawal is a proven remediation technology and at a relatively low cost to accelerate recovery in stratified lakes with high internal loading. Low dissolved oxygen (DO) content of discharged water has historically been a major problem with deep-discharge impoundments. However, multiple and shallower outlets have been incorporated into reservoir design to counteract low DO discharge.

In Europe, many Austrian lakes were mentioned in the literature after the construction of the Olszewski tubes, including Reither See and Hechtsee. Reither See, a dimictic pool, enhanced in quality after the construction of a pipe positioned at the full depth (8.2 m) in 1972. The tube diameter was 10 cm and water discharged at 0.24 m^3/min from the hypolimnion of the artificially formed pool, which is 1.5 hectares in area [42]. Epilimnetic TP decreased from annual means of at least 38 and 43 μg/L in 1974 and 1975 to at least 21 μg/L in 1977. Transparency rise by 40% during the four years after construction, and water clarity improved by ten times [42].

A larger pipe (18 cm) was installed in Hechtsee in 1973. The tube was positioned at a depth of 25 m (56.6 m full depth of the lake) to prevent the recreational area around the lake from odor contamination. The 26.3-hectare lake discharge ranged from 1.2 to 1.8 m3/min. For the first four years after construction, the P production (203 kg) surpassed the input (93 kg) by 110 kg, resulting in a real decrease in the TP content of the lake. TP above 24.5 m decreased by 70–80% t from 1973 to 1977, and TP below 25 m of the shift was marginal [42].

In a similar study, it was reported that the hypolimnetic withdrawal (HW) should conform to specific water-level targets throughout the growing season at the Pine Lake, Alberta, Canada. Long-term climatic conditions unveiled that water quantities for withdrawal were estimated to be adequate for seven to 10 years, and HW was feasible for the period range of 4.5 out of seven years in 1996–2005. Besides, the profile quantities of TP and total dissolved P were declined under a decreasing hypolimnetic anoxia phenomenon. Moreover, a 40% decrease in chlorophyll-a average concentration was achieved during treatment, while the Secchi transparency studied through years has reached the guideline of 3 m in the HW. In parallel fewer and less extensive cyanobacterial blooms were reported. The aforementioned advantageous water-quality improvements were achievable despite the certainly unchanged external and internal loads in the long-term [43].

There are several more recent applications such as in Poland and Italy. The Pławniowice Reservoir (Poland) is a former sandpit (created in 1975) that became hyper-eutrophic after about 20 years so that an in-lake treatment was required. Hypolimnetic withdrawal was conducted via three pipes, and results for the first eight years of operation (2004–2011) were reported [45]. P export was 2.5–3 times larger than P input so that the P retention was negative, about −300% [45].

At a relevant study of an installed HW system in the Avigliana lakes at Lago Avigliana Grande and Lago Annone (Italy), the total TP export was 631 kg or 693 mg/m^2 (Avigliana Grande) during the recorded period 2005–2010. Such an export should lead to a decreased in the epilimnetic TP growing period at 47%. At the Lago Annone lake, except for the year 2008, the observed treatment of water-column average TP concentration at fall turnover was decreased by 33%. The most pronounced decrease for the period of 2008–2013 was 51%. Moreover, for the period of 5–7 years, the TP export was more than six times of surface outflow, while for all years it was achieved an average increase of 1.5–13 times [43].

However, this technique has some drawbacks. This is brought up in a study done by Schauser and Chorus [46] who evaluated both external and internal control measures in two Berlin lakes. In Schlachtensee, the comparison between the P content of the lake, the external P load, and the P content withdrawn from the hypolimnion showed that, for the most part, hypolimnion withdrawal had only a minor effect on the total P budget of the lake and the phytoplankton and cyanobacteria biomass. The main source of P in the hypolimnion was redox-sensitive P desorption from sediments. However, stable thermal stratification during the summer is largely maintained by the desorbed P in the hypolimnion with minimal transfer to the epilimnion and is not directly accessible for phytoplankton growth [46].

Furthermore, the discharge of hypolimnetic water containing high concentrations of P, NH_4, H_2S, reduced metals and low oxygen may cause water-quality problems downstream (Fig. 4.10). If outflow streams contain important fisheries and are otherwise used for recreation or water supply, then special precautions are necessary to minimize adverse effects. For example, withdrawal water from Lakes Wononscopomuc and Waramaug is aerated and mechanically cleaned before being discharged downstream [42].

In revisiting the aforementioned literature production and insights on hypolimnetic withdrawal, it can be noted that this method has generally decreased the internal P loading at the sediment–water interface, whereas cyanobacterial blooms have not always been controlled. Understanding the reasons for success, or not, it is crucial to take into consideration that unsuccessful cyanobacterial responses are mainly caused by (a) lack of external control, (b) TP concentrations at higher to P limitation threshold, and (c) water column stability decrease thus leading the entry of bottom nutrient-rich waters into the epilimnion [29]. The advantages and disadvantages of the method have been summarized and comparably depicted in Fig. 4.10.

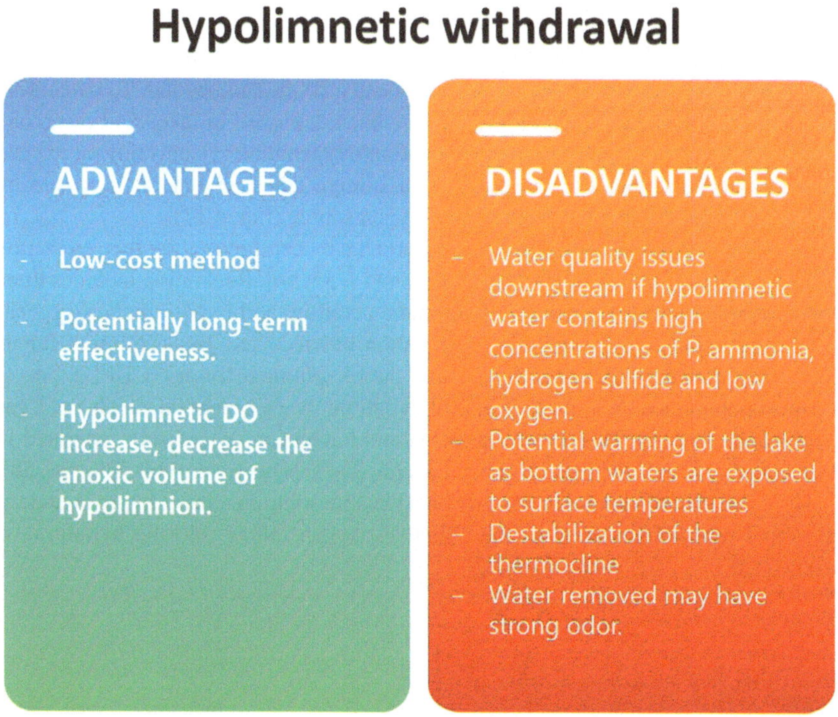

Fig. 4.10 Advantages and disadvantages of biomanipulation in shallow lakes. *Source* Author's own study

4.2.4 Reaeration

Anoxic conditions of deepwater are attributed to low dissolved oxygen, DO < 2 mg/L, caused by the degradation of organic matter (OM) in benthic water due to oxygen consumption, especially in cases where cut off of oxygen transfer is accompanying the formation of a thermocline in deep water systems [47–49]. An increase in internal phosphorus loading and elevation of primary production can be observed in anoxic conditions, thus, deteriorating water quality and depleting the supply of drinking water [50]. As the main countermeasure to this problem, the advancements of oxygen replenishment are the following: (a) phosphorus, iron, and manganese releases from the sediment are confined [29]; (b) nitrogen level reduction by stimulating bacterial growth [51]; (c) favorable living conditions for benthic organisms are nurtured [52] and (d) cost reduction of drinking water supply and mitigate of secondary pollution through by-products disinfection [53].

In the early years of hypolimnetic aeration, results were often variable, in part due to availability of pure oxygen, design and performance issues, and oxygen storage logistics. As research on hypolimnetic aeration systems continued, technology improved, hypolimnetic aeration has proven to be an effective technique with more predictive results [51]. However, costs for hypolimnetic aeration treatment remain high. Most hypolimnetic aeration applications, often in tandem with other restoration activities, have dated to the early 2000s and beyond. To date, more than 30 hypolimnetic aeration systems have been deployed in lakes and reservoirs worldwide [54]. Modern hypolimnetic aeration/oxygenation systems rely on oxygen stored adjacent to the lake or reservoir in storage tanks or generated on-site by pressure swing adsorption (PSA) as the oxygen source (Fig. 4.11).

Over the last few decades, research efforts have been focused on the restoration of hypolimnetic water systems via reaeration [53, 55, 56]. Reaeration includes three main categories: water-mixing, hypolimnetic aeration, and hypolimnetic oxygenation. Among these, hypolimnetic oxygenation has received great attention for its higher restorative efficiency [57]. Circular bubble plume diffusers (CBPD), one of the most popular technologies because of hypolimnetic oxygenation, consist of a gas source and a reaeration device (Fig. 4.11) and have been successfully operated in deep water systems but have never been operated in sub-deep reservoirs [53, 56, 58].

Hypolimnetic oxygenation has proven a feasible solution to treat hypoxia in deep lakes and reservoirs. However, the effective oxygen inflow into the hypolimnion

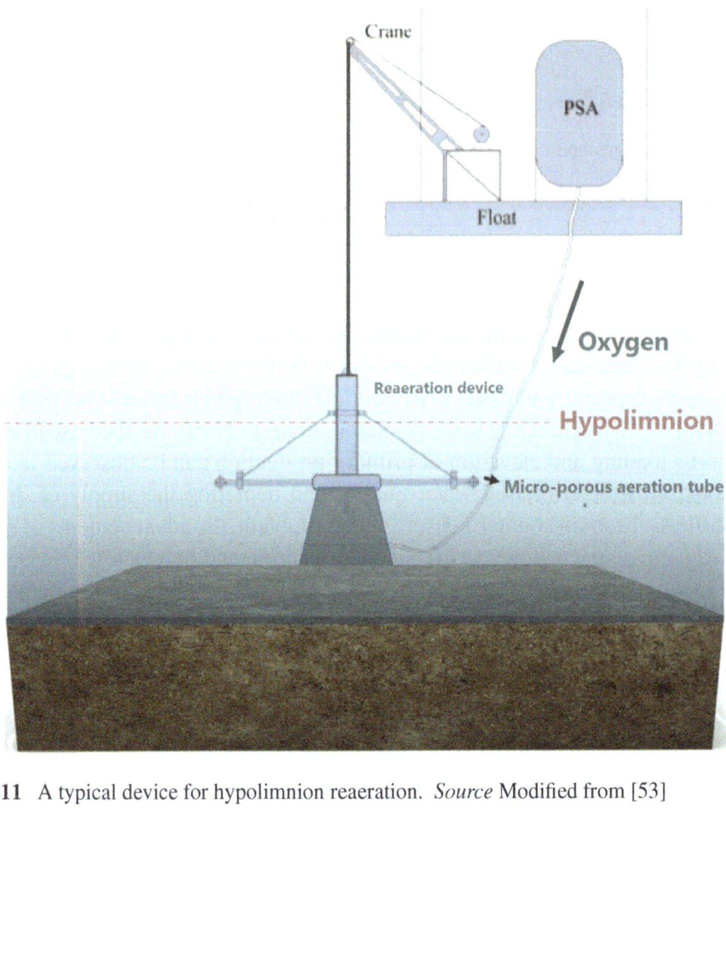

Fig. 4.11 A typical device for hypolimnion reaeration. *Source* Modified from [53]

at shallow lakes and a poorly designed hypolimnetic oxygenation system are linked to higher turbidity, weak thermal stratification, and sediments warming [54]. Contrarily, there is little knowledge accumulated regarding the viability of hypolimnetic oxygenation in shallow water sources. Gerling et al. [54] developed a successful test of hypolimnetic oxygenation in a shallow reservoir (Falling Creek Reservoir, USA), comparing its outcomes to previous studies. It was reported that the oxygenation system can increase the concentrations of hypolimnetic dissolved oxygen at a rate of ~1 mg/L/week without weakening stratification or warming the sediments. Moreover, the proposed deployment suppressed the release of reduced iron, manganese, and phosphorus, from the sediments [54].

Moore et al. [59] stated a successful application of long-term hypolimnetic aeration on phytoplankton and cyanobacteria dynamics in Newman's Lake (USA), a shallow eutrophic lake, where aeration was implemented since 1992. The authors confirmed that the system was capable of maintaining adequate oxygen to reduce internal P loading but needed to stay in operation year-round. They reported a good correlation between phytoplankton biomass and TP concentrations with a significant decrease over time. However, they also mentioned that external loads were not well monitored, and therefore, it was difficult to assess the efficiency of the hypolimnetic oxygenation [59].

In a recent study, hypolimnetic aeration was used, to restore water quality in the Aha Reservoir, in Southwest China [53]. A total of 95% of the imported oxygen was dissolved, but most was consumed by organic matter and other reduced substances within the hypolimnion. The oxygen consumption of organic matter, NH_4^+, and remaining reduced materials accounted for 41.4–52.5, 25, and 13.3–24.4% of the total imported oxygen, respectively. Compared with the control sites, dissolved oxygen levels in the hypolimnion increased 3–4 times, and concentrations of NH_4^+, total Fe, and total Mn were reduced by 15.5%, 45.5%, and 48.9%, respectively. A significant decrease in total phosphorus (TP) and nitrogen concentrations was observed in the experimental zone (0.04–0.02 mg/L and 1.9–1.7 mg/L, respectively). This indicates that aeration has great potential for oxygenation of the hypolimnion and improving water quality in the sub deep water system [53].

Also, maintaining the stability of thermal stratification is crucial to protect the living conditions for benthic organisms. As illustrated in Fig. 4.12c, there was a stable thermocline with a clear temperature gradient within the depth of 9–14 m before the experiment. During operation, the temperature of benthic water was the same as before and the thermocline remained in place (Fig. 4.12d). This indicates that the hypolimnetic oxygenation had no significant influence on the thermocline [53].

Oxygen dissolution efficiency (ODE) measurements indicated a high as 95% measured, which confirmed that most oxygen can be introduced into the aqueous phase. However, only moderated increases in DO concentration are achievable within the hypolimnion, except in the intrusion layer (Fig. 4.12a). The main quantities of oxygen introduced into the water were instantly consumed in the hypolimnion by algae and other reduced substances, including reduced Fe, Mn, sulfur, and NH_4^+, thus, resulting in the limited oxygenation effect. This finding was verified by the

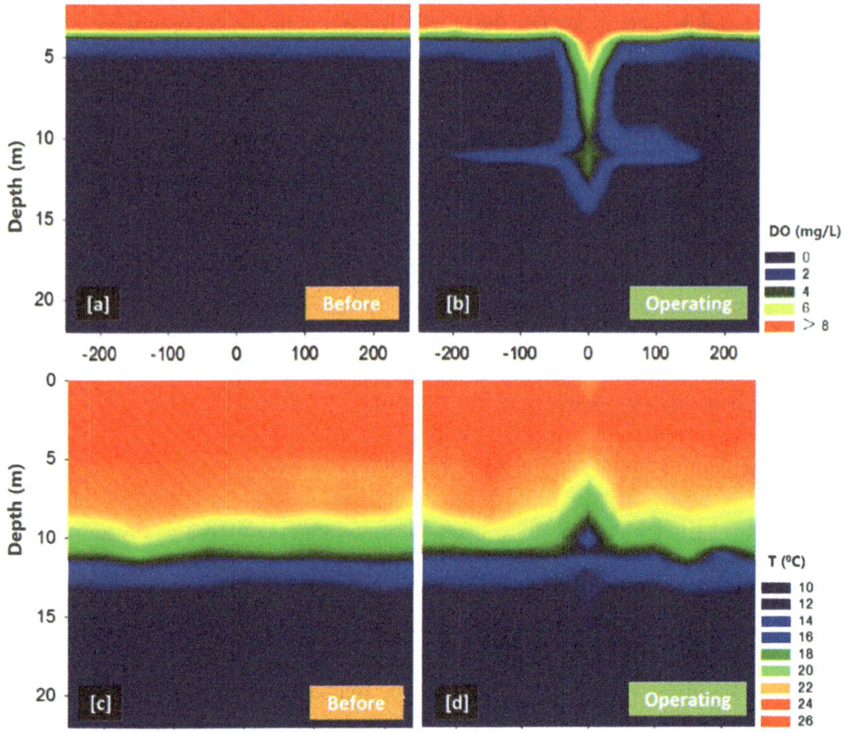

Fig. 4.12 **a** Spatial variation of dissolved oxygen (DO) in the water column before and during the application, and **b** Spatial variation of temperature in the water column before and during application. *Source* Modified from [53]

significant decline in NH_4^+, total Fe, and total Mn concentrations at the experimental zone, where it was measured net decreasing percentages of 15.5%, 45.5%, and 48.9%, comparing with the control sites, respectively [53].

Toffolon et al. [60] conducted a long-term analysis of 10 years, in studying the hypolimnetic oxygenation in Lake Serraia (Italy). These authors reported that although increased DO can be observed during operations, an increase in hypolimnetic temperature was probably due to the entrainment of hypolimnetic waters into the epilimnion. Similarly, a decrease in phytoplankton biomass and the replacement of cyanobacteria by chlorophytes were denoted. This analysis was explained by the reduction in P overall availability, due to the reduction in migrating the overall quantities of nutrients from the sediments [60].

However, one of the principal reasons is to maintain stratification during aeration preventing the recirculation of sediment-released P and maintain habitat for cold-water biota. If this does not occur, hypolimnetic aeration can be regarded as a less efficient restoration method. Low DO concentrations could also negatively affect fish communities, restricting fish movement.

Moreover, hypolimnetic aeration/oxygenation may not be effective if the water column is too shallow. Although stratification may exist, the density gradient may not be sufficient to resist thermocline erosion. In slow circulation conditions and if the water body destratifies, this may result in low DO throughout the water column and the introduction of toxic chemicals (e.g., hydrogen sulfide H_2S) into the epilimnion [61]. This would be of particular concern if epilimnetic water is being used as a drinking source. Hypolimnetic aeration is not recommended if the maximum depth is less than 12–15 m and/or the hypolimnetic volume is relatively small. While hypolimnetic aeration may restore oxygen conditions for fish and other biota, other toxic elements may not be sufficiently reduced to allow survival [61].

4.2.5 P-Inactivation Agents

Restoration has been defined as the return of a disturbed system to some pre-disturbance state that is both self-regulating and integrated into the larger landscape [62–65]. The chemical restoration methods mainly aim at reducing the P release from the sediment by improving the P binding capacity and thus creating P limitation of the phytoplankton.

Adding materials to lakes to adsorb dissolved P is a well-established method for improving water quality in eutrophic lakes. Phosphorus adsorption has been used worldwide since at least the late 1960s, most notably in the USA [61, 66]. The method involves applying materials to lakes to adsorb SRP and therefore reduce the bioavailability of phosphorus in the lake [67]. Materials often have an additional flocculating property, which causes the removal of suspended sediments from the water column that contain particulate nutrients. The most frequently used products have been aluminum salts, notably aluminum sulphate ('Alum'). Alum reduces internal phosphorus loading due to adsorption of SRP to aluminum hydroxide to form a stable complex that is unavailable for phytoplankton growth (Hickey and Gibbs 2009):

$$Al_2\,(SO_4)_{3x\mathrm{H2O}} + 2PO_4^{2-} \rightarrow 2AlPO_4 \downarrow 3SO_4^{2-} + xH_2O$$

In the last decade of research, studies have considered a plethora of different solid adsorbents as suitable materials to reduce P levels entering water bodies [21, 68–72]. Among the materials tested there were that of iron oxides, red mud, fly ash, zirconium oxide, and carbonates [73–78]. These materials contained plentiful advantageous applications but, also sustaining, certain disadvantages. It is noteworthy that iron oxide and red mud use for P reduction is predominately affected by the solution pH (Fig. 4.13) [78, 79]. Besides, stoichiometries are, in the majority of cases/experimental sessions, more than 1:1 regarding the reacting species with the phosphate anion. Therefore, the most important drawback of these adsorbents is the release of the adsorbed phosphate while changing specific chemical properties of the solutions tested, such as pH and redox change.

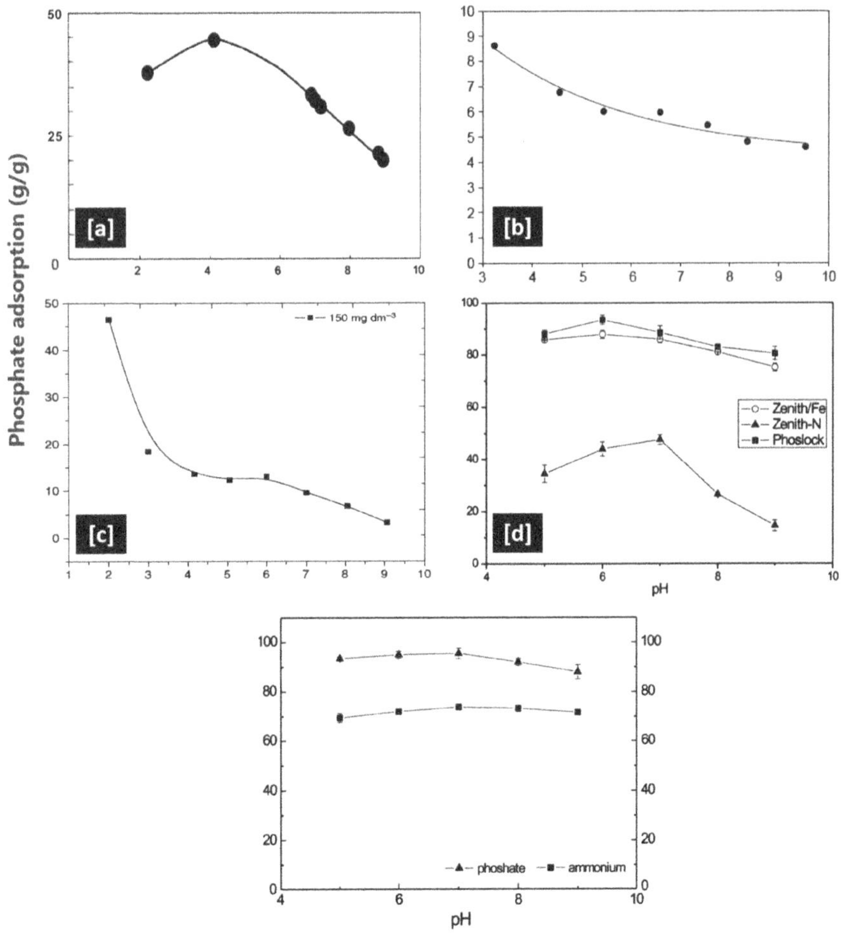

Fig. 4.13 Amount of phosphate adsorbed onto adsorbents at different Ph **a** aluminum oxide hydroxide, **b** red mud, **c** hydrous Zirconium oxide, **d** Zenith/N, Zenith/Fe and Phoslock™, and **e** Bephos™. *Source* Modified from [78–83]

Today, Alum, Calcite, Phoslock™ (lanthanum modified clay) and Modified Zeolite (Z2G1) are proven the most promising materials, being commonly used to restore eutrophic water bodies; however, the last two of them are considered the most innovative materials of lakes' remediation.

Materials are typically sprayed onto the lake surface with the objective of achieving an even application, although phosphorus adsorbing materials can also be used to bind phosphorus from external sources, e.g., by deploying in agricultural drains [84]. It is important to consider site-specific constraints when identifying appropriate products and application strategies. In particular, pH is an important consideration due to potential metal toxicity issues. For alum applications, it is necessary to buffer the lake to maintain pH > 6.5 to avoid the formation of highly toxic Al^{3+} ions. For lanthanum applications, studies to date have shown that concentrations generally remain below the acute toxicological threshold of different aquatic organisms, except for very low alkalinity waters [85].

4.2.5.1 Alum

Aluminum salts are common flocculation agents for a long time in water treatment and lakes' restoration. Among flocculation agents, alum or aluminum sulfate ($Al_2(SO_4)_3$) has been proven the best flocculant used in Europe and the USA [67]. Its functionality is based on the fact that Al hydroxides presented in lake sediments, like iron hydroxides, are natural P adsorbents, and since iron-bound P is soluble in reducing sediments, then the Al hydroxides are P binding preferred as a long-lasting sink for bioavailable P. Therefore, aluminum is acting as a barrier system of capturing and inactivating those potentially mobile P from the lake bottom sediments [14].

Alum is at its optimal pH values of alum performance is ranged at 6–8, while at these conditions it can effectively precipitate suspended particulates and, to a lesser extend, and to remove through absorbing soluble forms of phosphorus fractions. Alum is also able to absorb the dissolved reactive phosphorus (DRP) at anoxic conditions, thus the resulting floc to settle on the bottom of the sediment [14]. The application of alum to lake water in acidic solution necessitates its, foremost, buffering usually by sodium bicarbonate in order to keep pH more than 6.5. In case of low alkalinity of water, then it will occur below pH 6 dropping and the forming of highly toxic Al^{+3} species. This fact is critical to the facilitation of the flocculation process. At high pH water values, flocculation cannot be achieved, thus requiring the use of a low concentration alum solution. Another constraining parameter of disrupting the floc formation is the surface mixing effect through wind [86].

While adding alum to the water column, then alkalinity reduction is reported through the formation of aluminum hydroxide. Consequently, the exact determination of the maximum dosage of the added material is linked to the maximum added amount of Al up to a pH 6 value of lake water to be measured. In such a process, the suiting dose ensures the optimal maximum formation of the $[Al(OH)_3]$-P floc with the minimal possible forming of the toxic Al^{+3}. Calculating alum dose could be also based on the internal phosphorus among from the sediment, being calculated as five times [87–89].

Threshold Development for Factors Relating to Treatment Longevity

According to Huser et al. [90], 114 lakes treated with aluminum (Al) salts to reduce internal phosphorus (P) loading were analyzed to identify factors driving longevity of post-treatment water quality improvements. Control factors of the longevity of TP reduction and identification of critical thresholds can be achieved through the development of a decision tree partition modeling a sample of N = 68 lakes. The first threshold contained a dose of 5.1 g Al m^{-2}; thus at higher dose values, the mean treatment longevity for TP improvement was approximately eight years. From this group of lakes having Osgood Index (OI) higher than 5.7 [91] corresponds to a mean treatment longevity of 15 years [90]. Then, the reported threshold was that of Watershed to lake area ratios (WA:LA) of 8.8. In lower ratios, the longevity treatment of lakes increased to 26 years. In this group, there are lakes containing an Al dose of greater than 28 g Al m^{-2} and the corresponding average longevity of 36 years. The other side of the partition tree infers that at Al dose lower than 15.1 g Al m^{-2} and a WA:LA ratio greater than 27, these lakes sustained a reduced TP concentration less than one year.

4.2.5.2 Allophane

Allophane is a naturally occurring clay mineral that has a high aluminum and iron content, and as a consequence, a high affinity for phosphorus and is used as a capping agent. Its phosphorus binding ability was tested by Yuan and Wu [92] who documented its ability to bind phosphorus at water to sediment ratios similar to those found in some lakes. In these tests, the ability of allophane to bind phosphorus increased with increasing phosphorus concentrations. Similarly, Gibbs et al. [93] demonstrated that at environmentally relevant P concentrations, allophane would bind 5 g P kg^{-1} but could bind up to 17 g P kg^{-1} at higher concentrations of dissolved phosphorus. The major advantage that allophane has over the other products is that it is completely natural and therefore may be perceived by stakeholders as desirable and its use may therefore attract less restrictions than other geo-engineered products

4.2.5.3 Z2g1^{tm}

Aqual-P or Z2G1, as it was previously called, is a zeolite-based capping agent that has been modified with aluminum salts to provide active P binding (Hickey and Gibbs 2009). It was developed by Scion, Rotorua, and is manufactured by Blue Pacific Minerals Ltd., Matamata, New Zealand. Being an aluminum-modified zeolite, Aqual-P has the potential to bind both anions and cations giving it the ability to bind both PO_4^{3-} and NH_4^+. Previous work by Gibbs and Özkundakci [94] illustrated this ability, with Aqual-P.

4.2.5.4 Phoslock^{TM}

Phoslock^{TM} is La-modified bentonite clay being applied by the Australian Federal Government-owned Commonwealth Scientific and Industrial Research Organization (CSIRO) and the Australian Water and Rivers Commission to redress P fluxes generating eutrophic conditions in the Swan River Estuary, Western Australia [95]. P removal mechanism through Phoslock^{TM} involves a phosphate anion—La reaction resulting in the formation of a single insoluble species of lanthanum phosphate or rabdophane [86]. Phoslock™ is proven capable at a wider range of pH than other P-inactive methods, since it precipitates polyphosphates as well as orthophosphates, while the produced La-phosphate sustains very low solubility [96]. Phoslock™ is also less flocculent than alum, allowing for more effective additions in shallow lakes [97, 98].

4.3 Cost

The main objection to dredging in many cases is the high cost. Project-to-project cost comparison for sediment removal is difficult due to a large number of variables that affect dredging costs. These variables include equipment, project size or volume of material removed, disposal site availability, density of the material being removed, distance to the disposal area, and use of the removed material. Generally, the per-unit volume cost of sediment removal is inversely related to the total volume of material removed [42]. Data assembled by Peterson [99] from 64 sediment removal projects in the United States indicate a cost range from \$0.46/m^3 to \$27/m^3. Blazquez et al. [100] estimated costs between \$51 and \$2,127/m^3 for the removal of sediments (2020 USD). Dredging was reviewed for Twin Lake, Minnesota at a total cost of \$2,570,000 [101]. Removing 15 cm of sediment would have yielded approx. 13,500 m^3 of dredged material for a total capital cost of \$1.7 million. However, due to space limitations surrounding the lake, on-site disposal of the dredged material and high cost determined the treatment to be unfeasible (Table 4.1).

Minimal cost analysis for biomanipulation has been developed. A feasibility study conducted for Twin Lake, (USA) reviewed the cost of biomanipulation [101]. Due to fish migration between Sweeney and Twin Lake, the fish population was planned to be monitored at an annual cost of \$1,120 per year (USD, 2020). Four stocking events were determined to be included, which amounted to \$270,000 [101].

Relatively low capital and annual operational cost are advantages of hypolimnetic withdrawal (Table 4.1). Construction costs were estimated at \$400,000, and annual operation and maintenance cost (including electricity, chemical cost, and settle flocculent disposal) was estimated at \$40,000. Engineering and design cost was estimated at 20% of total capital costs. For a 20-year treatment life, the total cost of the project would total \$1,330,000 [101].

For hypolimnetic aeration, using the cost efficiency and the average airflow reported for the 15 aeration projects, which is 15.3 m^3/min or 3.7 kg O_2/min, amounts to approximately \$340,000 for 160 days per year for installation and operation cost. Based on the average area of the 15 lakes, the annual cost is approx. \$4,400/ha per year (2020 USD). Moreover, the cost of a new aeration system in the 1.3 km^2 Lake Nieuwe Meer (the Netherlands) was valued at around €750,000 per km^2 with annual operational costs of about €25,000 [102].

Table 4.1 Operating cost of each restoration method

Restoration method	Operating cost
Dredging	\$34–\$1,409 per m^3 [100] Twin Lake (USA): Total cost \$2,570,000 [101]
Biomanipulation	Inexpensive relative to other techniques. \$273,000 Twin Lake (USA), total cost, 20-year lifetime). \$127,000 Lake Nokomis (USA), total cost, 10-year lifetime) [101]
Hypolimnetic withdrawal	The advantage of this method is relatively low cost & potentially long-term effectiveness. Costs range around \$400 k and \$10 k per treatment [86]. Twin Lake, MN (8 ha): \$1.3 million (total cost, 20-year treatment life) [101]
Hypolimnetic aeration/oxygenation	\$4,400/ha per year (2020 USD)[a] A new aeration system in the 1.3 km^2 Lake Nieuwe Meer (the Netherlands) was valued at around €750,000 per km^2 [102]
P-inactivation agents	Phoslock: \$40–\$880 per k for bio-available P immobilized. \$0.75. (50 mg mobile P/kg in top 10 cm) \$6.02 (400 mg mobile P/kg in top 10 cm)

[a]2002 USD cost approx. \$3,000/ha per year (based on mean areas of 15 lakes) [42]

4.4 Conclusions

Based on the aforementioned analysis, it is noteworthy the dynamic evolution of environmental, seasonal, climatic, weather, and anthropocentric activities that all impacting on the lakes' well conditioning and long-term preservation. In this context, the following key aspects of the analysis have been creatively approached.

Sediment dredging is a short-term, extremely high cost (of removal, disposal and treatment) restoration method and therefore only practically applicable in small and shallow systems. While it directly removes the cyanobacterial inoculum and reduces the labile P content of the sediment, it can have negative impacts on benthic organisms. Biomanipulation usually involves the removal of rough fish such as carp and the introduction of piscivores (for example, bass and northern pike). Small planktivorous fish are grazing algae thus reducing zooplankton populations at lakes of low piscivores' populations. Besides, in mesotrophic lakes having moderate algae populations, such algae can be reduced even though zooplankton cannot entirely control the excessive production of algae. Algae grazing by zooplankton can effectively reduce algae population since the added piscivores can reduce the planktivorous fishes. This method is not invasive, being characterized as "natural", low cost, fishery improving, and moderating algal populations. However, even though this method is related to zooplankton increases and its heavy grazing, no complete algae reduction is reported. Subsequently, such a complete reduction of algae implies that the water quality goals by regulatory directives, such as that of Water Framework Directive, cannot be met by the sole criterion of surface water P concentrations' measured.

Hypolimnetic withdrawal is a measurable indicator of the direct removal of P-laden lake bottom waters, being especially fitting to control ecological impacts that are particularly benefited both zooplankton and fish habitats. Hypolimnetic withdrawal is advantageously preventing the P-accumulation in bottom waters and the migration of high P-quantities from bottom waters to the surface. Contrarily, the shortcomings of hypolimnic withdrawal are that of the following: (a) irritant odor and nuisance of the residential areas caused by water removal from the hypolimnion, thus, a summertime (of high internal loading) shut down is a social demand, (b) hypolimnetic withdrawal followed by treatment and discharge back to the lake is considered a costly process of high capital and operation demands, thus proven an inefficient approach to P-removal comparing to a whole-lake treatment, and (c) the thermal exposure of bottom waters to surface temperature is causing the lake waters' warming during treatment. In avoiding lake mixing, the warmed water has to be fed back to the appropriately designed depth of the lake column. Besides, whenever complete destratification is not advisable, the hypolimnetic aeration and oxygenation are proven as of short-termed effectiveness method at deep stratified systems; implying high operating costs at continuous modes of operation.

If the sediments contain high levels of P-binding chemical elements, mainly Ca, Mg, Al, that resist the effects of low oxygen, then even at low oxygen, P remains at the sediment bottom of the lakes. At higher P quantities of lake bottom waters when dissolved oxygen is low, P is primarily bounded on Fe instead of Ca, Mg, and Al. Consequently, the P-immobilization in the lake sediment necessitates the addition of the aforementioned three elements. Among them, Al is the most readily P-binding and it displaces Fe at lake sediment pH conditions. Calcium can also P-binding at lake sediment. However, this behaviour has been observed as an experimental approach, while it is proven more complicated and more expensive method. Therefore increasing use of P-inactivation materials (mainly PhoslockTM) has been reported, being proven a widely recognized restoration method to bind excess P in lake sediments while not affected by low oxygen levels in the lake bottom waters. Such a sediment P-inactivation is increasing dissolved oxygen in the hypolimnion since after treatment the algal growth at the lake is lower and less organic matter is deposited on sediments, all of which can result in oxygen depletion. However, P-inactivation has the constraint of repetitive application in the future. These issues are systematically discussed in Chap. 6. Conclusively, it is noteworthy that each method sustains strengths and weaknesses that are linked to each time system considered, thus, a combination of restoration methods applied is proven more successful than a single one. Since internal P-release is likely to increase from already-experienced severe climatic changes, then, the control methods should include internal P-loading increase in their applicability and effectiveness anticipated.

References

1. Smith VH (2009) Eutrophication. In: Encyclopedia of Inland Waters. ISBN 9780123706263
2. Selman M, Greenhalgh S (2010) Eutrophication: sources and drivers of nutrient pollution. Renew Resour J
3. Carpenter SR, Caraco NF, Correll DL, Howarth RW, Sharpley AN, Smith VH (1998) Nonpoint pollution of surface waters with phosphorus and nitrogen. Ecol Appl
4. Smolders AJP, Lamers LPM, Lucassen, ECHET, Van Der Velde G, Roelofs JGM (2006) Internal eutrophication: how it works and what to do about it—a review. Chem Ecol
5. Zamparas M, Zacharias I (2014) Restoration of eutrophic freshwater by managing internal nutrient loads. A review. Sci Total Environ
6. Sharpley A, Jarvie HP, Buda A, May L, Spears B, Kleinman P (2013) Phosphorus legacy: overcoming the effects of past management practices to mitigate future water quality impairment. J Environ Qual
7. Spears BM, Lürling M, Yasseri S, Castro-Castellon AT, Gibbs M, Meis S, McDonald C, McIntosh J, Sleep D, Van Oosterhout F (2013) Lake responses following lanthanum-modified bentonite clay (Phoslock®) application: An analysis of water column lanthanum data from 16 case study lakes. Water Res
8. Lee HW, Lee YS, Kim J, Lim KJ, Choi JH (2019) Contribution of internal nutrients loading on the water quality of a reservoir. Water (Switzerland)
9. Waters KA, Curran JC (2015) Linking bed morphology changes of two sediment mixtures to sediment transport predictions in unsteady flows. Water Resour Res
10. Curran JC, Waters KA, Cannatelli KM (2015) Real time measurements of sediment transport and bed morphology during channel altering flow and sediment transport events. Geomorphology
11. Burger DF, Hamilton DP, Pilditch CA (2008) Modelling the relative importance of internal and external nutrient loads on water column nutrient concentrations and phytoplankton biomass in a shallow polymictic lake. Ecol Modell
12. Song K, Adams CJ, Burgin AJ (2017) Relative importance of external and internal phosphorus loadings on affecting lake water quality in agricultural landscapes. Ecol Eng
13. Parsons CT, Rezanezhad F, O'Connell DW, Van Cappellen P (2017) Sediment phosphorus speciation and mobility under dynamic redox conditions. Biogeosciences
14. Jeppesen E, Søndergaard M, Jensen JP, Havens KE, Anneville O, Carvalho L, Coveney MF, Deneke R, Dokulil MT, Foy B et al (2005) Lake responses to reduced nutrient loading—an analysis of contemporary long-term data from 35 case studies. Freshw Biol
15. Callaghan D, Orihel D, Paterson M, Schindler D (2011) Importance of internal phosphorus loading in a hypereutrophic prairie lake [Killarney Lake, Manitoba] and its implications for lake remediation. Aquat Toxic Work
16. Gross C, Hagy JD (2017) Attributes of successful actions to restore lakes and estuaries degraded by nutrient pollution. J Environ Manage
17. Carpenter SR, Ludwig D, Brock WA (1999) Management of eutrophication for lakes subject to potentially irreversible change. Ecol Appl
18. Osgood RA (2017) Inadequacy of best management practices for restoring eutrophic lakes in the united states: guidance for policy and practice. Inl Waters
19. van Oosterhout F, Waajen G, Yasseri S, Manzi Marinho M, Pessoa Noyma N, Mucci M, Douglas G, Lürling M (2020) Lanthanum in water, sediment, macrophytes and chironomid larvae following application of Lanthanum modified bentonite to lake Rauwbraken (The Netherlands). Sci Total Environ
20. Zamparas M, Kyriakopoulos GL, Kapsalis VC, Drosos M, Kalavrouziotis IK (2019) Application of novel composite materials as sediment capping agents: column experiments and modelling. Desalin Water Treat
21. Waajen G, van Oosterhout F, Douglas G, Lürling M (2016) Management of eutrophication in Lake De Kuil (The Netherlands) using combined flocculant—Lanthanum modified bentonite treatment. Water Res

22. Zamparas M, Drosos M, Kalavrouziotis IK (2019) Sorption of phosphate from innovative composite material focusing on physicochemical interactions. Desalin. WATER Treat 151:212–218
23. Nürnberg GK (2007) Lake responses to long-term hypolimnetic withdrawal treatments. Lake Reserv Manag
24. Mehner T, Benndorf J, Kasprzak P, Koschel R (2002) Biomanipulation of lake ecosystems: successful applications and expanding complexity in the underlying science. Proc Freshw Biol
25. Lürling M, Waajen G, Engels B, van Oosterhout F (2017) Effects of dredging and lanthanum-modified clay on water quality variables in an enclosure study in a hypertrophic pond. Water (Switzerland)
26. Waajen G, van Oosterhout F, Douglas G, Lürling M (2016) Geo-engineering experiments in two urban ponds to control eutrophication. Water Res
27. Marotta H, Bento L, De Esteves FA, Enrich-Prast A (2009) Whole ecosystem evidence of eutrophication enhancement by wetland dredging in a shallow Tropical Lake. Estuaries Coasts
28. Jing LD, Wu CX, Liu JT, Wang HG, Ao HY (2013) The effects of dredging on nitrogen balance in sediment-water microcosms and implications to dredging projects. Ecol Eng
29. Bormans M, Maršálek B, Jančula D (2016) Controlling internal phosphorus loading in lakes by physical methods to reduce cyanobacterial blooms: a review. Aquat Ecolv
30. Liu C, Zhong J, Wang J, Zhang L, Fan C (2016) Fifteen-year study of environmental dredging effect on variation of nitrogen and phosphorus exchange across the sediment-water interface of an urban lake. Environ Pollut
31. Yu J, Ding S, Zhong J, Fan C, Chen Q, Yin H, Zhang L, Zhang Y (2017) Evaluation of simulated dredging to control internal phosphorus release from sediments: focused on phosphorus transfer and resupply across the sediment-water interface. Sci Total Environ
32. Cao X, Song C, Li Q, Zhou Y (2007) Dredging effects on P status and phytoplankton density and composition during winter and spring in Lake Taihu, China. Proc Hydrobiol
33. Zhong JC, Yu JH, Zheng XL, Wen SL, Liu DH, Fan CX (2018) Effects of dredging season on sediment properties and nutrient fluxes across the sediment-water interface in Meiliang Bay of Lake Taihu, China. Water (Switzerland)
34. Zhang S, Zhou Q, Xu D, Lin J, Cheng S, Wu Z (2010) Effects of sediment dredging on water quality and zooplankton community structure in a shallow of eutrophic lake. J Environ Sci
35. Penning H, Sørensen SR, Meyer AH, Aamand J, Elsner M (2010) C, N, and H Isotope fractionation of the herbicide Isoproturon reflects different microbial transformation pathways. Environ Sci Technol
36. Tsujimura S, Okubo T (2003) Development of Anabaena blooms in a small reservoir with dense sediment akinete population, with special reference to temperature and irradiance. J Plankton Res
37. Ruley JE, Rusch KA (2002) An assessment of long-term post-restoration water quality trends in a shallow, subtropical, urban hypereutrophic lake. Ecol Eng
38. Hickey CW, Gibbs MM (2009) Lake sediment phosphorus release management—decision support and risk assessment framework. New Zeal J Mar Freshw Res 43:819–856
39. Søndergaard M, Lauridsen TL, Johansson LS, Jeppesen E (2017) Repeated fish removal to restore lakes: Case study of lake væng, Denmark-two biomanipulations during 30 years of monitoring. Water (Switzerland)
40. Gianni A, Zamparas M, Papadas IT, Kehayias G, Deligiannakis Y, Zacharias I (2013) Monitoring and modeling of metal concentration distributions in anoxic basins: Aitoliko Lagoon, Greece. Aquat Geochem 19:77–95
41. Dunalska J (2003) Impact of limited water flow in a pipeline on the thermal and oxygen conditions in a lake restored by hypolimnetic withdrawal method. Polish J Environ Stud
42. Cooke DG, Welch EB, Peterson SA, Nichols SA (2005) Restoration and management of lakes and reservoirs, 3rd edn. CRC Press, Boca Raton. ISBN 0-7484-0772-3
43. Nürnberg GK (2020) Hypolimnetic withdrawal as a lake restoration technique: determination of feasibility and continued benefits. Hydrobiologia

44. Nürnberg GK (2017) Attempted management of cyanobacteria by Phoslock (lanthanum-modified clay) in Canadian lakes: water quality results and predictions. Lake Reserv Manag
45. Kostecki M, Suschka J (2013) The successful results of pławniowice reservoir (Upper Silesia Region—South of Poland) restoration by hypolimnetic withdrawal. Arch Environ Prot
46. Schauser I, Chorus I (2007) Assessment of internal and external lake restoration measures for two Berlin lakes. Lake Reserv Manag
47. Fortino K, Whalen SC, Johnson CR (2014) Relationships between lake transparency, thermocline depth, and sediment oxygen demand in Arctic lakes. Inl Waters
48. Zhang L, Li QH, Huang GJ, Ou T, Li Y, Wu D, Zhou QL, Gao TJ (2015) Seasonal stratification and eutrophication characteristics of a deep reservoir, longtan reservoir in subtropical area of China. Huanjing Kexue/Environmental Sci
49. Sánchez-Leal RF, Bellanco MJ, Naranjo C, García-Lafuente J, González-Pola C (2020) On the seasonality of waters below the seasonal thermocline in the Gulf of Cádiz. Cont Shelf Res
50. Perron T, Chételat J, Gunn J, Beisner BE, Amyot M (2014) Effects of experimental thermocline and oxycline deepening on methylmercury bioaccumulation in a Canadian Shield lake. Environ Sci Technol
51. Singleton VL, Little JC (2006) Designing hypolimnetic aeration and oxygenation systems—a review. Environ Sci Technol
52. Mehner T, Diekmann M, Gonsiorczyk T, Kasprzak P, Koschel R, Krienitz L, Rumpf M, Schulz M, Wauer G (2008) Rapid recovery from eutrophication of a stratified lake by disruption of internal nutrient load. Ecosystems 11:1142–1156
53. Lan C, Chen J, Wang J, Guo J, Yu J, Yu P, Yang H, Liu Y (2017) Application of circular bubble plume diffusers to restore water quality in a sub-deep reservoir. Int J Environ Res Public Health
54. Gerling AB, Browne RG, Gantzer PA, Mobley MH, Little JC, Carey CC (2014) First report of the successful operation of a side stream supersaturation hypolimnetic oxygenation system in a eutrophic, shallow reservoir. Water Res
55. Mobley M, Gantzer P, Benskin P, Hannoun I, McMahon S, Austin D, Scharf R (2019) Hypolimnetic oxygenation of water supply reservoirs using bubble plume diffusers. Lake Reserv Manag
56. Mcdougall TJ (1978) Bubble plumes in stratified environments. J Fluid Mech
57. McGinnis DF, Lorke A, Wüest A, Stöckli A, Little JC (2004) Interaction between a bubble plume and the near field in a stratified lake. Water Resour Res
58. Singleton VL, Little JC (2006) Designing hypolimnetic aeration and oxygenation systems—a review. Environ Sci Technol
59. Moore BC, Cross BK, Beutel M, Dent S, Preece E, Swanson M (2012) Newman Lake restoration: a case study part III. Hypolimnetic oxygenation. Lake Reserv Manag
60. Toffolon M, Serafini M (2013) Effects of artificial hypolimnetic oxygenation in a shallow lake. Part 2: numerical modelling. J Environ Manage
61. Cooke GD (2007) History of eutrophic lake rehabilitation in North America with arguments for including social sciences in the paradigm. Lake Reserv Manag
62. National Research Council Transportation Research Board HCM (2000) Highway capacity manual. In: Freeway facilities, adverse weather capacity reduction. ISBN 9780309160773 0309160774 9780309160780 0309160782 9780309160797 0309160790 9780309160803 0309160804
63. Shapiro J, Lamarra V, Lynch M (1975) Biomanipulation: an ecosystem approach to lake restoration. Limnol Res Cent
64. Hupfer M, Hilt S (2008) Lake restoration. Encycl Ecol Five-Volume Set. ISBN 9780080914565
65. Martin J (2013) Restoration and management of lakes and reservoirs. Hydro-Environ Anal
66. Welch EB, Degasperi CL, Spyridakis DE, Belnick TJ (1988) Internal phosphorus loading and alum effectiveness in shallow lakes. Lake Reserv Manag
67. Cooke GD, Welch EB, Peterson S, Nichols SA (2016) Restoration and management of lakes and reservoirs

68. Lin J, Zhong Y, Fan H, Song C, Yu C, Gao Y, Xiong X, Wu C, Liu J (2017) Chemical treatment of contaminated sediment for phosphorus control and subsequent effects on ammonia-oxidizing and ammonia-denitrifying microorganisms and on submerged macrophyte revegetation. Environ Sci Pollut Res
69. Li X, Chen J, Zhang Z, Kuang Y, Yang R, Wu D (2020) Interactions of phosphate and dissolved organic carbon with lanthanum modified bentonite: implications for the inactivation of phosphorus in lakes. Water Res
70. Zamparas M, Kyriakopoulos GL, Drosos M, Kapsalis VC, Kalavrouziotis IK (2020) Novel composite materials for lake restoration: a new approach impacting on ecology and circular economy. Sustainability
71. Wang C, Wu Y, Wang Y, Bai L, Jiang H, Yu J (2018) Lanthanum-modified drinking water treatment residue for initial rapid and long-term equilibrium phosphorus immobilization to control eutrophication. Water Res
72. Lürling M, Tolman Y (2010) Effects of lanthanum and lanthanum-modified clay on growth, survival and reproduction of Daphnia magna. Water Res 44:309–319
73. Kharissova OV, Dias HVR, Kharisov BI, Pérez BO, Pérez VMJ (2013) The greener synthesis of nanoparticles. Trends Biotechnol
74. Summers RN, Smirk DD, Karafilis D (1996) Phosphorus retention and leachates from sandy soil amended with bauxite residue (red mud). Aust J Soil Res
75. Dursun S, Guclu D, Bas M (2006) Phosphate removal by using activated red mud from Seydisehir Aluminium Factory in Turkey. J Int Environ Appl Sci
76. Liu D, Wang P, Wei G, Dong W, Hui F (2013) Removal of algal blooms from freshwater by the coagulation-magnetic separation method. Environ Sci Pollut Res
77. Dhanke P, Patil A, Kore V, Thakare P, Patil U, Wagh S (2018) Phosphate removal from industrial wastewater effluent using modificoal fly ash. Desalin Water Treat
78. Rodrigues LA, Maschio LJ, Coppio LDSC, Thim GP, Da Silva MLCP (2012) Adsorption of phosphate from aqueous solution by hydrous zirconium oxide. Environ Technol (United Kingdom)
79. Tanada S, Kabayama M, Kawasaki N, Sakiyama T, Nakamura T, Araki M, Tamura T (2003) Removal of phosphate by aluminum oxide hydroxide. J Colloid Interface Sci 257:135–140
80. Zeng L, Li X, Liu J (2004) Adsorptive removal of phosphate from aqueous solutions using iron oxide tailings. Water Res 38:1318–1326
81. Zamparas M, Drosos M, Georgiou Y, Deligiannakis Y, Zacharias I (2013) A novel bentonite-humic acid composite material Bephos™ for removal of phosphate and ammonium from eutrophic waters. Chem Eng J 225:43–51
82. Zamparas M, Gianni A, Stathi P, Deligiannakis Y, Zacharias I (2012) Removal of phosphate from natural waters using innovative modified bentonites. Appl Clay Sci
83. Zhi Y, Zhang C, Hjorth R, Baun A, Duckworth OW, Call DF, Knappe DRU, Jones JL, Grieger K (2020) Emerging lanthanum (III)-containing materials for phosphate removal from water: a review towards future developments. Environ Int
84. Wagner KJ (2017) Preface: advances in phosphorus inactivation. Lake Reserv Manag
85. Copetti D, Finsterle K, Marziali L, Stefani F, Tartari G, Douglas G, Reitzel K, Spears BM, Winfield IJ, Crosa G et al (2015) Eutrophication management in surface waters using lanthanum modified bentonite: a review. Water Res
86. Hickey CW, Gibbs MM (2009) Lake sediment phosphorus release management-Decision support and risk assessment framework. New Zeal J Mar Freshw Res
87. Huser BJ (2017) Aluminum application to restore water quality in eutrophic lakes: maximizing binding efficiency between aluminum and phosphorus. Lake Reserv Manag
88. Huser BJ, Futter M, Lee JT, Perniel M (2016) In-lake measures for phosphorus control: The most feasible and cost-effective solution for long-term management of water quality in urban lakes. Water Res
89. Augustyniak R, Grochowska J, Łopata M, Parszuto K, Tandyrak R, Tunowski J (2019) Sorption properties of the bottom sediment of a lake restored by phosphorus inactivation method 15 years after the termination of lake restoration procedures. Water (Switzerland)

90. Huser BJ, Egemose S, Harper H, Hupfer M, Jensen H, Pilgrim KM, Reitzel K, Rydin E, Futter M (2016) Longevity and effectiveness of aluminum addition to reduce sediment phosphorus release and restore lake water quality. Water Res
91. James WF, Sorge PW, Garrison PJ (2015) Managing internal phosphorus loading and vertical entrainment in a weakly stratified eutrophic lake. Lake Reserv Manag
92. Yuan G, Wu L (2007) Allophane nanoclay for the removal of phosphorus in water and wastewater. Sci Technol Adv Mater
93. Gibbs MM, Hickey CW, Özkundakci D (2011) Sustainability assessment and comparison of efficacy of four P-inactivation agents for managing internal phosphorus loads in lakes: sediment incubations. Hydrobiologia
94. Özkundakci D, Hamilton DP, Gibbs MM (2011) Hypolimnetic phosphorus and nitrogen dynamics in a small, eutrophic lake with a seasonally anoxic hypolimnion. Hydrobiologia
95. Ross G, Haghseresht F, Cloete TE (2008) The effect of pH and anoxia on the performance of Phoslock®, a phosphorus binding clay. Harmful Algae
96. Zamparas M, Gavriil G, Coutelieris FA, Zacharias I (2015) A theoretical and experimental study on the P-adsorption capacity of Phoslock^{TM}. Appl Surf Sci 335:147–152
97. Vopel K, Gibbs M, Hickey CW, Quinn J (2008) Modification of sediment–water solute exchange by sediment-capping materials: effects on O 2 and pH. Mar Freshw Res 59:1101
98. Gibbs MM, Hickey CW, Özkundakci D (2010) Sustainability assessment and comparison of efficacy of four P-inactivation agents for managing internal phosphorus loads in lakes: sediment incubations. Hydrobiologia 658:253–275
99. Peterson SA (1982) lake restoration by sediment removal. JAWRA J Am Water Resour Assoc
100. Blazquez CA, Adams TM, Keillor P (2001) Optimization of mechanical dredging operations for sediment remediation. J Waterw Port Coastal Ocean Eng
101. Barr Feasibility Report for Water Quality Improvements in Twin Lake Crystal • Golden Valley • Medicine Lake • Minneapolis Minnetonka • New Hope • Plymouth • Robbinsdale • St. Louis Park Feasibility Report for Water Quality Improvements in Twin Lake CIP Projec. 2013
102. Lürling M, Mucci M (2020) Mitigating eutrophication nuisance: in-lake measures are becoming inevitable in eutrophic waters in the Netherlands. Hydrobiologia

Chapter 5
Chemical Lake Restoration Methods: From Alum to Innovative Composite Materials

Grigorios L. Kyriakopoulos, Miltiadis G. Zamparas, Xiaolei Sun, Meng Li, and Marios Drosos

Abstract This chapter evaluated a spectrum of chemical restoration approaches of eutrophic water systems, highlighting the key role of internal nutrient loads to guide remediation methods that control eutrophication. The usage of P- and joint P-/N-inactivation agents have been considered as feasible restoration tools through capacity and application methods in alignment with a wide spectrum of individually-examined nature-contexts, including that of water quality, algal blooms, flora, and fauna. Moreover, an integrated approach was developed to evaluate those restoration tools of degraded waters through specified conditions when P-inactivation agent was applied.

Keywords Lake restoration · Alum · Phoslock^{TM} · Calcite · Bephos^{TM} · Field studies

G. L. Kyriakopoulos
Photometry Laboratory, Electric Power Division, School of Electrical and Computer Engineering, National Technical University of Athens, 15780 Athens, Greece

M. G. Zamparas
Laboratory of Sustainable Waste Technology Management, School of Science and Technology, Hellenic Open University, Parodos Aristotelous 18, 26335 Patras, Greece

X. Sun
Institute of Bio- and Geosciences, Agrosphere (IBG-3), Forschungszentrum, 52425 Jülich, Germany

M. Li
Institute of Wetland Research, Chinese Academy of Forestry, Beijing 100091, China

M. Drosos (✉)
Institute of Resource, Ecosystem, and Environment of Agriculture (IREEA), Nanjing Agricultural University, 1 Weigang Road, Nanjing 210095, China
e-mail: drososmarios@njau.edu.cn

M. G. Zamparas and G. L. Kyriakopoulos (eds.), *Chemical Lake Restoration*,
https://doi.org/10.1007/978-3-030-76380-0_5

5.1 Introduction

Phosphorus and nitrogen have been recognized in literature as nutrients that primarily determined the harmful algal growth in eutrophic waters. The main inflows of external pollutants from wastewater and runoff waters as well as the embedded sediment-bonded nutrients, they are rebounding to the water column, being released in particularly intensive hypoxic conditions at sediment–water interfaces. Therefore, it is anticipated that strategies of integrated restoration can be certainly considered towards controlling both external and internal nutrient loads (Fig. 5.1). In valuing the potential risks involved to localized biota and, then, selecting the appropriate restoration technologies, it is important for researchers to conduct a background site investigation and to comprehend those effects of pollutant migration (Fig. 5.1). Therefore, it is of utmost importance for long-term monitoring plan and evaluation of the final restoration, while better assessing and feasibly ensuring the success of such a treatment.

Based on this assumption, it is very important to investigate the phosphorus (P) sediment forms and the relationships between physicochemical parameters of a field water body before treatment [1]. This study aims to enhance the knowledge of P cycling, thus, better controlling that water quality and the trophic status, highlighting the priority to emphasize in eutrophic waters management (Fig. 5.1).

Fig. 5.1 Integrated approach of lake restoration. *Source* Authors' own study

Point or non-point sources (external inputs) are intercepting the entrance of nutrients at lakes, thus, playing a vital role for eutrophication control. In an integrated approach, it is significant to examine the effectiveness of nutrients' removal, through nutrients' quantities and resultant water quality in the littoral zone of riparian buffers along with the watershed around eutrophic lakes. Overall, the effective removal of the total phosphorus (TP) and the total nitrogen (TN) can be achieved through the construction of wetland or grass/forest and grass riparian buffer strips, respectively [2, 3].

Several restoration methods have been used to remediate contaminated sediment which includes sediment dredging [4], hypolimnetic oxygenation [5], and sediment capping [6]. However, the release of dredged sediment pollutants can markedly disturb sediment texture or, even, function as secondary source of contamination to overlying water. Also, dredged sediment can occupy a large area of land, and its removal is expensive. Moreover, numerous researches stressed out that in controlling sediment nutrients release, oxygenation is not always proven as an effective method [7–9]. Contrarily, recent research attractiveness has shown that the reduction of dissolved P in the water and the prevention of P release from the sediment be achieved by using P-inactivation agents [10–12].

The use of Alum is a very widespread restoration method and has taken place in many ecosystems with different results in terms of water quality. In this context, it is critical to consider those conditions of functioning of Alum, as capable restoration tool in eutrophic waters. In the last decade, Phoslock^{TM} (lanthanum-modified bentonite) is gradually replacing Alum with its applications in aquatic ecosystems suffering from eutrophication. This chapter will analyze the two P-inactivation agents that are the two most commonly used worldwide.

5.2 Alum

Aluminum salts are common flocculation materials for a long time in water treatment and inland water restoration. Among the flocculation agents, alum or aluminum sulfate ($Al_2(SO_4)_3$) is the most widely used capping agent in Europe and the USA [13]. As mentioned in the previous chapter (Chap. 4), its functionality is based on the fact that Al hydroxides present in lake sediments, like iron hydroxides, are natural P adsorbents; thus, they bind P acting as a long-lasting sink for mobile phosphorus. Therefore, aluminum is acting as a barrier system of holding and inactivating those bioavailable P from the lake bottom sediments [14].

Aluminum salts, mainly as aluminum sulfate (Alum), can be introduced to the water column, forming $AlPO_4$ (aluminum phosphate) and a colloidal $Al(OH)_3$ aluminum hydroxide floc to which certain P forms bind. The aluminum hydroxide floc settles to the sediments where it still adsorbs and retains P within the lattice of the molecule, even under anoxic environments [15]. Alum application is dating back to more than 200 years, being the most well-known and widespread tool used for drinking water treatment worldwide [16]. Given the growing usage of aluminum additions to reduce eutrophication, the solubility and chemistry of aluminum are not well known [15]. However, alum is usually not suited for small, vigorous lakes since it creates flocs and enhances turbidity [17].

Optimal pH values of alum performance range from 6 to 8, where it can effectively precipitate suspended particulates and, to a smaller extent, remove through absorbing soluble forms of phosphorus (Fig. 5.2). Alum is also able to absorb the dissolved reactive phosphorus (DRP) at anoxic conditions, thus the resulting floc settles at the bottom of the sediment [14]. The application of alum to lake water in acidic solution necessitates its, foremost, buffering usually by sodium bicarbonate to keep pH more than 6.5. Consequently, lake water of low alkalinity, can drop pH lower than 6, leading to the formation of highly toxic Al^{+3} species. This fact is critical to the facilitation of the flocculation process. However, flocculation is not achievable at high values of water pH, thus, low concentration alum solution is required. Another

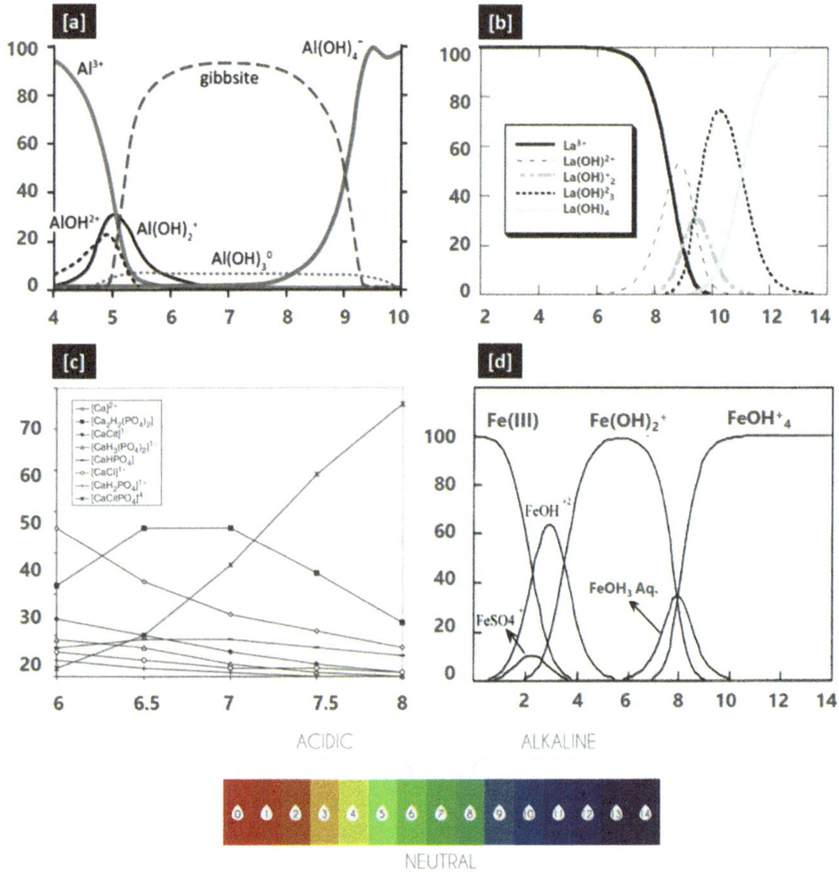

Fig. 5.2 Speciation diagram of **a** Alum, **b** Lanthanum, **c** Calcite, and **d** Fe(III). *Source* Modified from [22–24]

constraining parameter of disrupting the floc formation is the surface mixing effect through wind [18].

Due to Alum application in lake water as acidic solution, it has to be maintained at a pH value more than 6.5, thus, buffering with usually sodium bicarbonate is needed (Table 5.1). Consequently, in a lake water of low alkalinity, it can lead in pH drop lower than 6, and the formation of highly toxic Al^{+3} species will occur. The formation of toxic Al^{+3} species is mainly achieved by reverse hydrolysis of the $Al(OH)_3$ being toxic to the lake fish (Fig. 5.2). This step has to support the flocculation process. Indeed, in a lake water of high pH, flocculation cannot be achieved, while requiring the utilization of an alum solution of low concentration. The floc formation could be also impeded by wind currents due to the effect of surface mixing [17–19]. While Alum addition to water column is progressing, then, the alkalinity is lowered through the aluminum hydroxide formation. Therefore, the determination of alkalinity reduction in the water column can be calculated by the maximum amount of Al dosage added, until the pH of lake water be adjusted to 6. Such a targeted application dose enables that the optimum formation of the $[Al(OH)_3]$-P floc can be achieved by minimum occurrence of the toxic Al^{+3}. Alternatively, the calculation of alum dose can be based on the internal phosphorus amount from sediments, being estimated as five times the average internal phosphorus load at summer [20, 21].

5.2.1 Laboratory, Mesocosms, and Field Studies

In the study of Han et al. [25], microcosm experiments evaluated the effects of alum treatment on cyanobacteria-lysing organisms. It has been reported that at pH 6, the alum concentration is the maximum one where no toxicity is caused by monomeric Al. This concentration value is the maximum dose determined. Seven days post-treatment, there is release of intracellular microcystin-LR to the extracellular environment, reaching more than 95% while chlorophyll-a cannot be easily released from inside the cell. Alum treatment can damage cyanobacteria-lysing organisms and microcystin bacteria, as well as Microcystis cells. Subsequently, alum treatment with maximum dose (48 mg L^{-1}) is not suitable to remove cyanobacterial bloom without cyanotoxin release in reservoirs and ponds [25].

Moreover, non-target effect is caused by floc settling towards smothering of the lake bed (Table 5.1). Regarding sediment toxicity, no published data disclosed that it can be caused by alum, while information should focus on tolerance thresholds for sediment living species between different thicknesses of alum [26]. In earlier study, dating back to 1990, five alum-treated lakes reported an increase of faunal density and diversity in almost all of them [27]. In this context, Narf [27] assessed the changes in the accommodated profundal benthic and macrobenthic fauna among the aforementioned five eutrophic lakes, implying faunal density and diversity increases among these lakes' treatment. Chaoborus punctipennis (Say) and *Chironomus* spp. were the dominant organisms reported [27].

Table 5.1 Summary of innovative P-inactivation materials for reducing internal nutrient loads. *Source* Modified from [40]

Chemical method	Application	Mechanism	P adsorption capacity [mgP/g]	pH range[a]	Harmful effects
Phoslock™	Applied as a slurry or as granules	$La^{3+} + PO_4^{3-} \rightarrow LaPO_4$	10.5	5–8	No toxic effect to *Daphnia magna* by Phoslock™, because it is rapidly settled after application
Modified Zeolite	Applied as a slurry or as granules	$Al_2(SO4)_3 + 2PO_4^{-3} \rightarrow 2AlPO_4 \downarrow + 3SO_4^{-2}$	12.7	5–7	No discernable changes in plankton species composition when MZ is applied as a coarse-grained material that settled out of the water column rapidly
Alum	Applied as a liquid	Same as Alum $Al_2(SO4)_3 + 2PO_4^{-3} \rightarrow 2AlPO_4 \downarrow + 3SO_4^{-2}$	12–13.8	5.5–8.2	Lake water pH reduction-Acidification Toxicity Settling floc → smothering of the benthos
Calcite	Applied as a slurry	$3Ca^{2+} + 2PO_4^{-3} = Ca_3(PO_4)_2$	6.4	8.5	No concerns
Ferric chloride	Liquid to surface or sub-surface to hypolimnion	(I) $2FeCl_3 + 3Ca(HCO_3)_2 \rightarrow 2Fe(OH)_3 \downarrow + 3CaCl_2 + 6CO_2$ (II) $Fe^{3+} + PO_3^{3-} \rightarrow FePO_3 \downarrow$			No concerns

[a]pH range in which the material shows the maximum efficiency

In closely examining the adverse effects of lake inflow treatment with alum, laboratory and field investigations were performed at Fish Lake in Eagan, MN, and Tanners Lake in Oakdale, USA [28]. In such an approach, the designed tests showed that within 6 h an almost complete Al settling, whereas slight affection of final residual Al concentrations was reported by dose. The total Al entering Fish Lake has shown an average concentration of 346 and 458 μg Al L^{-1} at doses of 1 and 8 mg Al L^{-1}, as well as 70 and 152 μg L^{-1} at average in-lake concentrations. At the Tanners Lake, the inflows of average Al concentrations were higher, but—due to greater size—the Al concentration in the lake was lower. Benthic invertebrates at Fish Lake were protected in the lake at the settling pond, but at 8 mg Al L^{-1} treatment, it eliminated floc accumulation of almost all invertebrates in the pond. Moreover, no observed mortality effects (NOECs) were reported in literature due to Al concentration's occurrence. A measure of unbound and positively charged Al species had been compared to NOECs in order to estimate the potential aquatic toxicity from alum treatment. In this respect, at pH values of treated water entering higher than 6, then, only negligible potential risks of aquatic toxicity should be anticipated [28].

From the relevant literature, it has been stated that the careful consideration of pH and application rates should support the proper alum application. Subsequently, such a material can effectively sequester enough DRP while creating P limitation in the water. This can remediate eutrophication and eliminate the bloom of toxic cyanobacteria in favor of harmless phytoplanktonic species, as diatoms are [29]. Alum, except for its primary role as flocculant agent, it could be proven effective to support sediment capping [30]. Indeed, its application at the end of summer stratification is proposed, since there are anoxic conditions of the bottom lake water and maximum accumulation of DRP. Regarding the implementation and duration of alum applied, it has been proposed in the literature, the range of 5–20 years effectiveness before treatment repetition to be applied [31]. Nevertheless, no or low time effectiveness of alum treatment is anticipated, whenever a large portion of lake is covered by macrophytes, which during decomposition release high P-amounts into the water column [19].

In a similar study, Welch examined the affection of alum treatment among 21 lakes of U.S.A., focusing on this treatment effectiveness and longevity [32]. The determining parameters of effectiveness were that of lake TP reduction, the internal loading rate, and the chlorophyll-a. In 6 out of 9 polymictic lakes (that is two-thirds of the sample investigated) internal loading rate was reduced and lake TP was reduced by almost one-half for the period of 5–11 years. In parallel, at dimictic lakes an average of 80% reduction of internal loading rate was reported for the period of 4–21 years (averaged of 13 yrs). Among 6 polymictic lakes that showed an effective treatment, an average decrease of chlorophyll-a was initially reported, but this reduction was about 40% less than the pre-treatment level. Additionally, among 7 dimictic lakes an average of 57% initial reduction of chlorophyll-a was reported, followed by a 42% reduction after 5–18 years.

Huser et al. [17] investigated the internal P loading reduction through alum in 114 lakes, thus, identifying those factors of improving the post-treatment water quality. While identifying the key-aspects and the critical thresholds of achieving the longevity of TP reduction, a decision tree partition modeling in a sample of N = 68 lakes was developed. The threshold dose of 15.1 g Al m^{-2} was reported, while the mean treatment longevity for TP improvement at higher dose values was approximately 8 years (Fig. 5.3). Among those lakes having Osgood Index (OI) higher than 5.7, a mean treatment longevity of 15 years was reported [17]. Similarly, a ratio of Watershed to Lake Area (WA:LA) 8.8 was measured, whereas in lower ratios an increase to 26 years longevity treatment was reported. It is indicatively noted that lakes containing a higher than 28 g Al m^{-2} dose, the relevant average longevity reached that of 36 years. At the left-branch side of the partition tree, it can be inferred that at lower than 15.1 g Al m^{-2} dose and a higher than 27 WA:LA ratio, the relevant lakes sustain a TP-reduced concentration of less than 1 year.

There are a few notable cases of alum field applications.

While there are numerous studies devoted to short-term effects of alum field applications, there are few studies referring to long-term effects. Among such long-term studies, Lewandowski et al. [33] examined the annual treatment of the hardwater lake, Süsser See, Germany, with aluminum sulfate for the period 1977–1992. The authors conducted sediment-core-laboratory and in situ experiments of sediment and

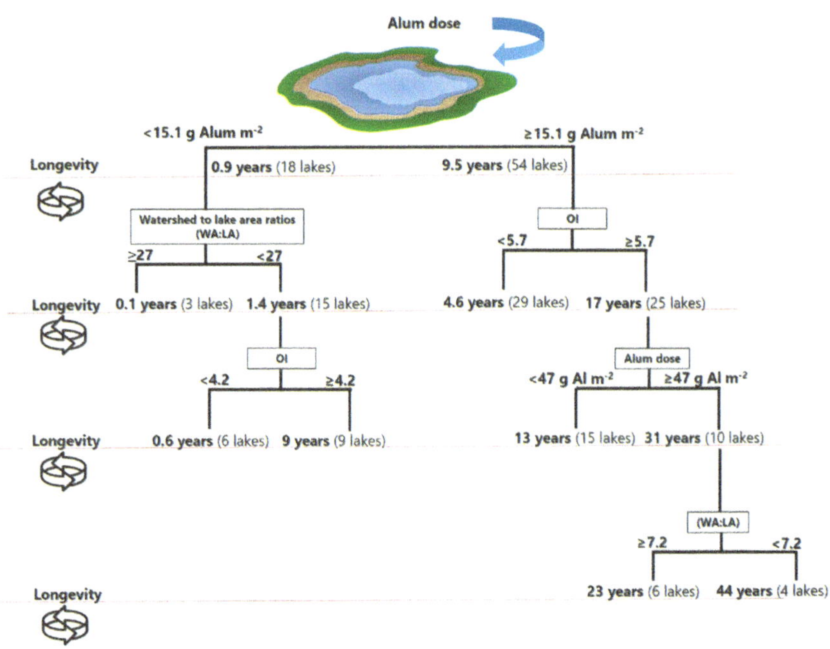

Fig. 5.3 Partition model with threshold values for the longevity of TP reductions concerning the number of lakes. The Osgood index (OI) can value the impact of lake morphology on treatment longevity. OI is representing the point of stratified or polymictic lakes, where the lower ratios of Watershed to Lake Area (WA:LA) are utilized to assess the impact of hydrology on treatment longevity. *Source* Modified from [17]

pore water. This treatment has identified the sediment depth profiles while increasing total P, NaOH-extractable P, and NaOH-extractable aluminum. Al has been added to bound excess P in the molar ratio of Al:P as 2.1:1. Regarding the in situ, in sediment core experiments and sorption batch experiments, the pore water profiles of soluble reactive P disclosed that the layer's sorption capacity of $Al(OH)_3$ is still not exhausted, enabling further P sorption at various P-fractions. The magnitude of the downward flux into the P sorbing layer is effecting on P sediment release. Besides, at both sediment core, modeling experiments have proven that the buried P sorbing layer is weakly affecting P release at the uppermost fresh sediment layers.

In a similar study, observations of the zooplankton community were made before, during, and for three years following the restoration of the eutrophic Medical Lake, USA, 1977, by applying 936 metric tons of alum in order to interrupt the lake's internal P cycle [34]. A reduction in the percent of rotifers has occurred and a similar rise in the crustaceans has been observed in reaction to the altered trophic position and co-occurring experiences within the community. Species richness indices showed a shift toward more diverse zooplankton populations with fewer yearly variability [34].

Earlier diagnostic studies with paleolimnological analyses and a detailed P budget at Lake Morey, Vermont, 1986, revealed the cause to be internal P loading from anoxic hypolimnetic sediments. The aluminum treatment can support sharp declines in total P and chlorophyll-a concentrations, and increased Secchi transparency. An 44 g Al/m^2 application can sufficiently form an effective barrier to sediment P release, further sustaining advantageous conditions to the next four years after application. On the other hand, the post-treatment conditions showed the elevated dissolved aluminum levels and adverse effects on the benthic invertebrates and the yellow perch population [35].

In this context, it is noteworthy that in the case of the oligomictic Newman lake in Washington State in 1989 with medium depth (30 m), having a stable thermal summer stratification beginning mid-to-late May and remains until mid-August or early September. In this case, there were two main alum applications: one whole lake application in 1989 and one microfloc in 1997. Annual average growing season TP ranged from 43 to 77 μg/L with an increasing trend before restoration. The first alum application lasted for 6 days in September 1989 using an aluminum dose of 2.8 mg/L [36]. In 1998, the year following the lake treatment, TP dropped to 33 μg/L, however in the next two years, phosphate concentration increased to pre-restoration levels, showing the limited effectiveness of the treatment. Aeration technique and alum application can cause limited treatment longevity, because increased oxygen demand along with phosphates' precipitation by alum, can heighten internal P loading while overwhelming the retention capability of the floc [36].

The key difference between alum microfloc installation (being applied at an operational phase at mid-summer of 1997) from conventional whole lake alum treatments is that these form large aluminumhydroxide flocculants, which because of their size settle rapidly and effectively remove particulate and solute P. The main research target is to inactivate P by chemically bonding it with the sediments [37]. This method has been successful in reducing nutrients and algae and in improving water clarity, offering more opportunities for $Al[OH]_3$ to react with particulate and dissolved phosphorus. On the other hand, for small lakes or small-scale implementations, alum injection may be sufficient in accomplishing water clarity on its own, as well as achieving algae and nutrient reduction goals, having the constraint of not being used as direct replacements for whole- and large-lake treatments, thus, not offering a long-term and flexible tool for lake managers. The successful evaluation of microfloc alum injection systems for lake restoration indicates further deployment needed [30, 37].

In a relevant study, the spraying of 13 m^3 of alum solution to the lake Okaro in December 2003 was applied by a boat that produced an 0.6 g/m^3 of Al in the epilimnion, aiming to study the impact of this material on the quantities of water nutrients and the trophic status indicator chlorophyll-a [38]. It was observed that total P decreases and chlorophyll-a increases, implying P flocculation by the alum addition, while phytoplankton may not be P-limited due to sufficient internal P load or limitation by N. Besides, low dose rate of the alum application is linked to low alkalinity of the lake Okaro, since there is a risk of striking pH decrease and alkalinity following the application. High pH values reduce the P sorption capacity, thus there is a persistence of soluble Al in water. Contrarily, the hypolimnion pH was below the lower threshold of pH 6 for adsorption. Following the phase of excessive dosing of soluble Al concentrations may endanger the protection of freshwater species for at least three months, though no fish deaths were observed [38]. After dosing and the Al adsorption capacity, it should be noticed among most nutrient species, their decrease in the hypolimnion, burial of the floc layer on the bottom sediments, or bioturbation. The alum dosing was also followed by a decrease in phosphate concentration in the epilimnion, while the dosing phosphate concentrations were low and represented only a small proportion of the total P pool. Besides, the main quantity of near surface P presented in phytoplankton was associated with the other, less easily adsorbed forms of organic particulate matter by the alum floc, compared to that of phosphate ions. Moreover, the post-dosing phase was followed by a substantial increase of ammonium in the epilimnion. An explanation of ammonium increase should be attributed to the mineralization of dead organisms, mainly zooplankton by heterotrophic bacteria and mortality of heterotrophic bacteria from alum dosing. It was particularly noted that the surface ammonium increased in alignment with phosphate decrease, the day after the dosing, possibly leading to a reduction in N occurrence of phytoplankton. In conclusion, the pre- and post-dosing situation made able the following conclusions reported: (a) no significant changes in water clarity, (b) hypolimnetic pH decreased to 5.5, (c) soluble aluminum exceeded the setting guidelines for freshwater organisms' protection, (d) epilimnetic phosphate decreased from 40 to 5 mg m^{-3}, (e) the mass ratio of TN:TP has been increased from 7:1 to 37:1, (f) on dosing application phase it was noted that the dominant phytoplankton species changed from *Anabaena* spp. before dosing, to *Ceratium hirudinella*, then *Staurastrum* sp. after dosing. It was also notable that the effectiveness of dosing may be limited by the sampling duration, design, and low alum dose. Moreover, the low alkalinity lake water and the coincidence of alum dosing with a cyanobacterial bloom and high pH are responsible for the occurrence of both hypolimnetic pH and epilimnetic TP decrease, and Al^{3+} and chlorophyll-a increase [38].

In better ameliorating a variety of eutrophication-related problems, it has been reported in the relevant literature a treatment of a 24 mg Al/l at the Green Lake, Washington, during the period March and April, 2004 [39]. The collection of four sediment cores six months after the alum treatment in Green Lake was conducted in order to explore the short-term effects of alum on sediment phosphorus speciation. Comparing the two ratios of Al to Al–P (112:1) to the average ratio of ~11:1 in other alum-treated Washington lakes, it had shown an approximately 10% of the binding capacity of the Al added. Such added Al supported binding potential capacity of a total of 21.6 g/m^2 P. Sediment inorganic P in the forms of Fe–P and labile-P can support internal loading to the concentration of 2.8 g/m^2. The application of the alum to the Green Lake enabled both the inactivation of the remaining inorganic mobile P, as well as the control of future P mobilization from the pool of organic sediment P [39].

Field measurements of whole-lake alum-treated eutrophic Spring Lake, during October and November 2005, were undertaken, and the data collected in the past two years 2003 and 2004 were compared. They unveiled a reduction of soluble reactive phosphorus (SRP) and total phosphorus (TP) concentrations in the water column the summer after the alum application, while no severe alteration was reported to both chlorophyll-a levels and irradiance profiles. Total macroinvertebrate density declined significantly in 2006 compared with 2004, with chaoborids and oligochaetes experiencing the greatest reductions. Internal P release rates calculated by using sediment cores incubated in laboratory anaerobic conditions, ranged from −0.052 to 0.877 mg TP m^{-2} d^{-1}. It is noteworthy that although the alum treatment effectively reduced internal P loading in Spring Lake, water column P concentrations remained at high concentrations in this system, possibly attributed to high external loading levels, and affected by the high chlorophyll levels. Therefore, the consideration of reducing internal and external inputs of P is necessary to a watershed managerial approach in order to address the cultural eutrophication of Spring Lake studied [41].

Treatment wetlands, 'in-pipe' measures, and in-lake aluminum sulfate (alum) treatment were applied to restore water quality in the four urban lakes [42]. Different alum dosing methods led to 4–20 (modeled) years of water quality improvements in the Minneapolis Chain of lakes in Minnesota (USA) after treatment, and only one of the four lakes continues to meet water quality goals approximately 25 years after the project started [42]. The overall external loads were reduced by an average of 13% at three reservoirs. Alterations to inner P sediment release rates after implementation of alum associated well with epilimnetic total P (TP) concentrations in these lakes, suggesting that improvements in water quality were mainly motivated by reduced internal loading by in-lake actions. In terms of dollars expended per unit of phosphorus extracted, the total volume of alum used in lake treatment was on average 50 times more efficient than in-catchment measures. While major external nutrient reductions might not be enough to sustain water quality in urban lakes, ongoing in-lake management of P deposited in lake sediment is both essential and cost-effective compared to in-catchment strategies [42].

In a recent study, Brattebo et al. [43] applied an alum treatment to the hypereutrophic Lake Ketchum improving to mesotrophic conditions after application in 2014 and 2015. From 2013 pre-treatment conditions, epilimnetic TP declined by 95%, that is from 289 to 34 μg/L in 2014, and to 15 μg/L in 2015. Hypolimnetic TP declined 99% overall, and chlorophyll-a dropped 70% while there was 135% improvement in transparency over the 2 years. Toxic cyanobacteria blooms, a chronic problem in the lake, disappeared in 2015. The water quality improvements were achieved even with the maintenance of a high (>450 μg/L) external input from an agricultural source. The changes were consistent with or exceeded predictions of a 2-layer, seasonal mass balance TP model used to evaluate restoration alternatives, implying that alum treatments can eliminate high internal loading and, while coupled with annual maintenance treatments, they can provide enduring and improved water quality despite continued external loading [43].

Based on the above long-term and thorough studies, it can be concluded that an incomplete pre- and post-treatment monitoring of nutrients' bioavailability can cause difficulty to determine the true and measurable effectiveness of P loading reduction to lakes. In other words, the managerial approach of estimating eutrophication risks by using total P concentrations is proven problematic since only a portion of P forms are biologically available for phytoplankton growth, whereas the TP measured in surface water, although indicating the water quality standards, cannot differentiate the P forms that reflect markedly different bioavailability. Such a TP indicator can support a detailed assessment of sources made via mass balance and seasonal accounting through monitoring programs that measure both external and treated flows and distinguish between bioavailable and recalcitrant nutrient forms. These multifaceted factors may affect the longevity of existing nutrient reduction measures aiming at improving the future restoration policies and planning (Fig. 5.4). In economic terms, the treatment cost is a significant aspect to make the option to reduce a particular nutrient source, internal or external, thus, promoting the most cost-effective treatment.

5.2.2 *Cost of Alum Treatment*

The study of Huser et al. [42] is indicative and particularly helpful in concluding treatment cost of Alum. These authors estimated a total restoration cost of $12,000,000 at a Minneapolis restoration project and a cost of $560,000 for all four alum sessions of treatment. The construction costs of more than 6 million USD did not include ongoing running costs for grit chambers and green infrastructure, such as constructed wetland and pond systems. It has to be also signified that there are notable, more than one order of magnitude, different costs' estimated between internal and external measures. However, in the light of the amount of money needed for per unit P removal, the lake alum treatment is, on average, approximately 50 times more effective than incatchment measures, which range from 44 (vs. constructed wetlands) up to more than 600 (vs. grit chambers and CDS units). All these techno-economic conditions, in alignment with the effectiveness of nutrients' reduction anticipated, they should

Fig. 5.4 Steps for maximizing Alum application effectiveness. *Source* Authors' own study

be taken into consideration for the selection of those management options that are mostly employed for long-term management of water quality [42].

To estimate the costs and evaluate the long-term effectiveness of an alum treatment, sediment cores were collected from eight sites in the lake to measure phosphorus that is mobile under anaerobic conditions (redox-P) to develop an alum dose for the lake. Sediments from each site were exposed to increasingly higher doses of alum in the laboratory until the measured mobile P (redox-P) was converted to aluminum-bound P (Fig. 5.5a). Two treatment zones were identified, prescribing 100 g Al/m^2 in areas greater than 15 feet deep and 50 g Al/m^2 in shallow areas demonstrating high mobile P in the sediment. This dose targeted a reduction in mobile phosphorus in the top 6 cm of sediments to control sediment phosphorus release. The cost for completing the alum treatment was estimated at just under $900,000.

Prior to investing in the alum treatment for Bald Eagle Lake, in-lake costs were compared to the cost of further watershed TP load reductions for other restoration projects (Fig. 5.5b). The 30-year life cycle cost for several types of watershed practices completed in the region was compared to the cost of an alum treatment on a per pound of TP removal basis (Fig. 5.5b). Diffuse watershed practices such as rain gardens were the most expensive since they had relatively low impact on phosphorus loading. As practices increased in size and adopted the use of adsorption media, such as iron-enhanced sand filters, cost efficiencies increased significantly. The five alum projects reviewed for this comparison were by far the most cost-effective projects,

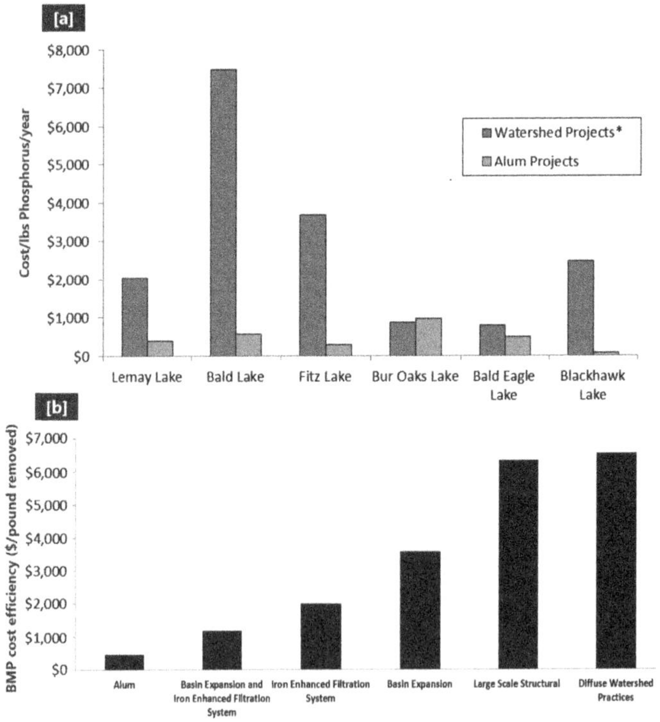

Fig. 5.5 **a** Phosphorus reduction Best Management Practice cost efficiency (Cost Efficiency = cost per pound/year), *Average cost of all potential watershed projects, **b** Best Management Practice (BMP) cost efficiency per project type

even though initial costs can be daunting. This analysis allayed the fears of spending such a large sum of money on the alum treatment as compared to spending more money on watershed practices.

Costs per kilogram P bound (or formed PAl) calculations of the new injection method indicated that the lakes of Flaten and Långsjön necessitated striking higher treatment costs compared to lakes in which Al has been applied to the water column (Table 5.2) [44]. Such an observation is reasonable, since there is need of extra working-hours for an injection treatment to be completed, while an overall treatment cost increase should be also caused by the use of buffered PAC compound. A research limitation is that the per kilogram P bound costs for (of higher costs) water treatment was not available. The injection treatment is related to binding efficiency increases, but the concurring cost differences can be offset by higher quantities of P bound per unit Al applied, implying that other than direct costs, measurable criteria can be considered [44]. Moreover, the short-term stress to biota in the water column and its potential for negative effect controlling are mainly related to short-term reduction of plankton due to physical settling. The aforementioned treatments may be less intensive due to lower possibilities of negative effects' occurrence, while potentially reducing treatment cost. Moreover, since alum contains sulfur that under anoxic conditions are forming iron sulfides, the Al compounds are possible to decrease the overall binding capacity of the sediment. Under the precondition of substantial reduction of productivity in treated lakes, then, improvement of oxygen conditions and binding increase by Fe can occur whenever alum is utilized [44].

Comparing Alum treatment with other methods, the following conclusions can be made: Aluminum treatment costs ranges at $500–$1,500/acre, being accompanied by decrease of the areal cost for larger treatments, unbuffered treatments, and less monitoring requirements. Higher costs are incurred by extreme controls and monitoring. It can be also signified that iron treatment and alum treatment costs are similar to each other, since the chemicals are of low cost to purchase but there is need of higher doses to be applied. However, the joint application of iron and aeration systems is recommended, thus, it is reported a substantial higher total project cost. From a chemicals' point of view, it is also noteworthy that calcium treatments are slightly less expensive than alum, but they are proven effective in hard water

Table 5.2 Dose, total amount of formed PAl, and treatment costs in different lakes worldwide

Lake	Area (ha)	Treated area (ha)	Dose (g/m^2)	P_{Al} formed (kg)	Total cost (thousands US$)	Cost per kg P_{Al} formed (US$)	References
Flaten	63	40	61	3660	353	152	[44]
Långsjön	29	29	75	1450	388	268	[44]
Harriet	139	45	32	7670	136	18	[45]
Cedar	70	60	27	1929	218	113	[42]
Calhoun	180	130	42	10920	241	22	[42]
Isles	44	26	18	955	109	114	[42]

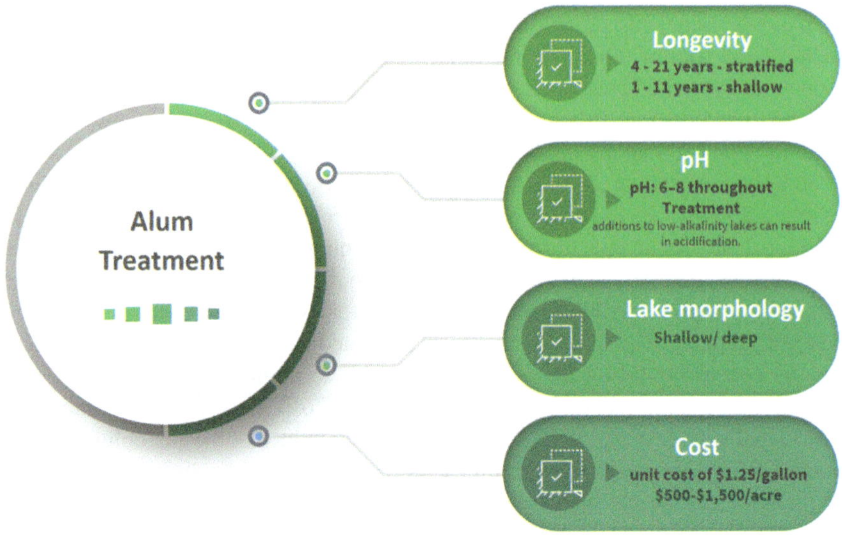

Fig. 5.6 Alum treatment summary. *Source* Authors' own study

lakes of high natural pH. Contrarily, in cases of neutral to acidic waters, the applied calcium remains dissolved, thus, it cannot precipitate and inactivate the targeted P (Fig. 5.6).

5.3 Phoslock^{TM} (Lanthanum Modified Bentonite, LMB)

Investigations conducted over three decades ago suggested that rare earth elements (REE), and in particular lanthanum (La), may be effective in reducing phosphate concentrations from lake water in pilot-scale studies, from wastewaters, and for the removal of straight-chain polyphosphates [46, 47]. This potential was established in later studies for both artificial and natural pond water with a good phosphate adsorption capacity, achieved at a wide pH range. Melnyk et al. [48] have reported that phosphate precipitation by lanthanum (a rare earth element) was more effective in a wide pH range (4.5–8.5) than either Fe or Al salts. The molar ratio of La to phosphate anion is 1:1 as resulting from the $La^{3+} + PO_4^{-3} \rightarrow LaPO_4$. The solubility product (K_{sp}) of the La-phosphate complex is particularly low (24.76 and 27.92) in both aqueous and seawater solutions, making it the least soluble of the rare earth–phosphate complexes [49]. It is assumed that the complex is too insoluble in low La and phosphate concentrations and low pH environments [50]. Additionally, with temperature increasing, the solubility of the La-phosphate decreased, thereby suggesting the strength of its bond [49]. Using GEOCHEM-PC computer diagrams, Diatloff et al. [50] displayed that the complex can be formed in solutions with a pH as low as 4.0. Through their work, they established that the reaction of the two components can occur at pH 4 and with very low concentrations of La and phosphate species. Lanthanum is a rare earth element that is very potent while removing P from the water with a reported ratio of 1:1, comparing to sodium aluminate $NaAlO_2$ showing the inefficient ratio of 7:1 to achieve similar uptake of P [51, 52]. Moreover, the point of zero charge (PZC) of La oxides is higher than that of other known adsorbents [53, 54], implying its effectiveness at high pH rates (Fig. 5.7). In examining the role of Lanthanum to reduce phosphates, it is noted that its functionality is related to the concentration and the rate of application, being toxic to aquatic organizations [47]. Overcoming this constraint can be achieved by embedding La into clay minerals' structure, such as bentonite, while considering the advantageous feature of their cation exchange capacity (CEC).

Subsequently, a La-modified bentonite was developed in 90s by CSIRO (Commonwealth Scientific and Industrial Research Organization) [19, 55]. Phoslock™ (Lanthanum-modified bentonite, LMB) manufacturing is related to exchange of lanthanum ions with the bentonite internal surface, being randomly adsorbed as exchangeable cations.

Phoslock™ forms the rhabdophane ($LaPO_4$) mineral that is highly stable in the presence of P species such as the common oxyanions of orthophosphates [56]. This mineral is an analog of other REE-phosphate minerals that were demonstrated to be resistant to weathering over geological time. The phosphate uptake capacity of lanthanum-substituted clays typically exceeds 10 mg P/g [57]. Large-scale field trials of lanthanum-substituted clays suggest that they may be effective in capturing a

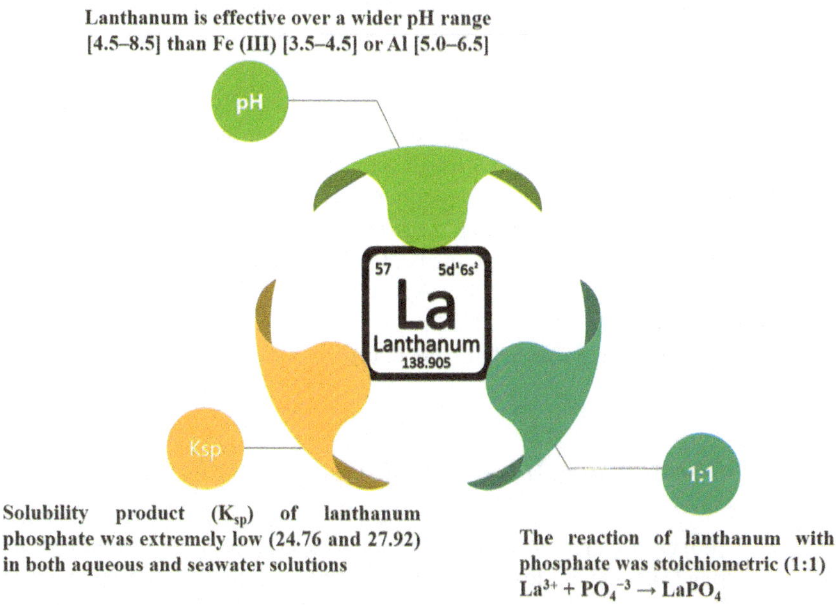

Fig. 5.7 Research outcomes from phosphate adsorption in Lanthanum

substantial proportion of phosphate released from bottom sediments when applied as a thin (1 mm) active capping [58, 59].

The rare earth element in the clay matrix will either interfere with the phosphate anion in the water body or stay in the clay mineral structure under a variety of environmental conditions [60]. Although La ions could be toxic to some aquatic organisms, particularly cladocerans such as *Daphnia*, hence the potential liberation of La ions from the bentonite meaning a significant environmental risk, Phoslock™ has been classified as not hazardous [61, 62]. Leachate experiments showed a La release, but at insignificant quantities of 0.13–2.13 μg/L, from the material. In such a case, Phoslock™ was applied in a worst-case dose of 250 mg/L [62]. In the relevant study of Geert et al. [61], these authors investigated the potential lanthanum accumulation/toxicity of Phoslock™ in animals and humans. In such a case, in experimented rats, lanthanum was administered orally, either in the forms of lanthanum carbonate, lanthanum chloride, or Phoslock™ in doses of 0.5 or 17 mg/L for ten weeks. It was reported that LMB use does not pose a toxicity risk, while the gastrointestinal absorption of lanthanum had been kept in minimum quantities that were related to the type of the compound. Besides, except for the liver, no significant increase in lanthanum levels was observed in the other organs studied while, based on previous studies, slightly increasing levels of liver lanthanum at the worst-case scenario did not indicate hepatotoxicity risks [61]. Moreover, in the study of Álvarez–Manzaneda et al. [63], the authors carried out tests to evaluate acute and chronic toxicological effects of LMB on *Raphidocelis subcapitata* and on *Daphnia magna*, where no inhibition patterns were observed in the case of the algal growth inhibition test through LMB [64]. In direct contact exposure to organisms, that was in the immobilization test of *D. magna*, the EC_{50} values disclosed no statistically significant differences [63].

Phoslock™ application to treated water is implemented into two application techniques: While spreading it in a granular form or as a thick suspension, then a solid/water slurry has been created through spray. In the phase of its settling through the water column, orthophosphates are permanently bounded and resting on the sediment, as capping material acting (Fig. 5.8c) to prevent P release from the sediment [64, 65]. It is estimated that for more than 250 lakes worldwide, the Phoslock™ has been applied (Fig. 5.8 and Table 5.4 at Appendix), this being the primary agent in reducing the sediment P release. The sizes of such treated lakes are varying from pond sizes up to 190 ha (Table 5.4 at Appendix).

5.3.1 Laboratory and Mesocosm Studies

Vopel et al. [66] used a microelectrode to calculate the evident gas flux of the capping layers resulting from various doses of LMB. The effects of these layers on pore water pH and dissolved oxygen (DO) were also tested [66]. From this study, it was revealed that capping material raised the depth of the oxic–anoxic interface and associated pH minimum and altered the sediment O_2 consumption as functions of the capping layer

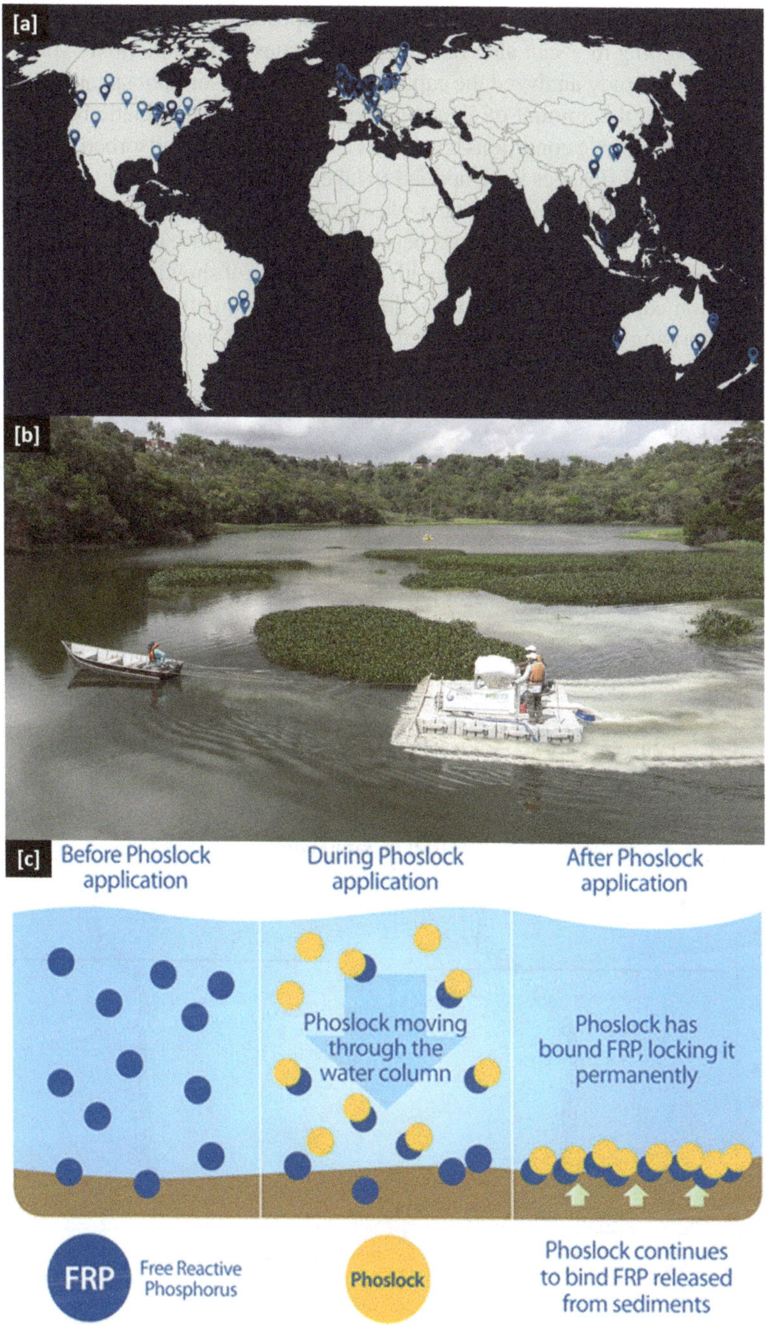

Fig. 5.8 **a** Applications of Phoslock™ worldwide, **b** injection of Phoslock™ slurry in the water, **c** How Phoslock™ works. Available at: https://www.phoslock.com.au/site/content/default.aspx

thickness and apparent diffusivity. Thus, it is proposed that capping layers derived from doses >200 g m^{-2} can alter benthic process rates and solute fluxes (Fig. 5.9).

A laboratory study analyzed the capacity of LMB to adsorb phosphates under the effect of pH and the amount of algae and how these affect its kinetic behavior and what impact the anoxic conditions have on its ability to hold the adsorbed phosphates. Also, higher LMB dosages were applied to lake water which had pH above 9 to see if it was possible to achieve a greater phosphorus removal [67]. It was found that LMB was most effective at pH values between 5 and 7 and with a pH increase from 7 to 9, phosphorus removal decreased with any pH above 9 showing large diminution in adsorption capacity. In comparison, the adsorption potential was smaller in lake water comprising algae than in prepared reverse osmosis water solutions. A crucial

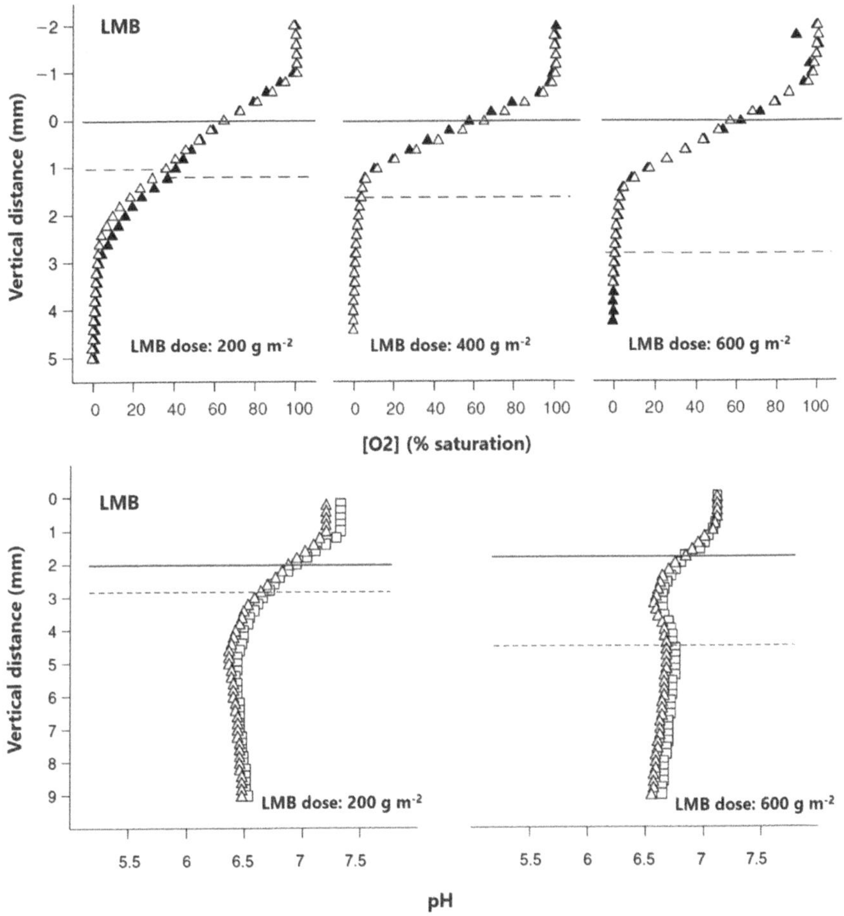

Fig. 5.9 Vertical O_2 microprofiles of lake sediment capped with LMB in different quantities (gm^{-2}). Solid (red) and dashed lines (grey) indicate approximate positions of the surfaces of the capping layers and the sediments respectively. *Source* Modified from [66]

issue is that P was not released from the LMB under anoxic conditions. Similar results related to the wide pH range were recorded from Zamparas et al. [57]. This is attributed to the La ions producing excessive hydroxyl ion concentrations and thereby reducing the P-binding sites on the surface of the LMB. The precise pH at which La hydroxides begin to precipitate is 8.35 [57, 68], so, subsequently, the adsorption capacity may be decreased above that pH. The sample of lake water which had pH 9, after applying the LMB had an increased rate of turbidity reduction than that with a pH value of 5. The particle size evidence confirms this inference. In water samples comprising algae, the adsorption potential was smaller than in synthetic water collected by reverse osmosis due to the existence of humic acids [68]. This result is supported by a more recent study [69] which shows that HAs contribute to phosphorus retention in lakes, presumably by complexing with phosphorus [69].

Ross and Cloete [67] presented that anoxic conditions did not show phosphorus release from the LMB. Some other concluding remarks from this study were: (a) the conductivity of all the water solutions and samples was not affected by LMB and (b) The reduced FRP (filterable reactive phosphorus) adsorption capacity observed at a pH above 9 could not be overcome by increasing the LMB dosage to a 450:1 treatment ratio [67]. The research goal of a new study by Zamparas et al. [10] was the estimation of LMB as a P-inactivation agent in the sediment–water interface under oxic and anoxic conditions from a polluted pond (Greece). LMB resulted in about ~81% and ~82% decrease of the phosphate flux from oxic and anoxic sediments respectively. Also, the impact of dissolved organic matter (DOM) on phosphate sorption capacity (PAC) of Phoslock™ was studied experimentally and theoretically, simulating natural eutrophic waters. By increasing the concentration of DOM, the sorption efficiency is dropped. However, it is noteworthy that in the case of LMB application, the PAC is retarded, whereas in the case of Z2G1 (was used for comparison with LMB) is sharp. In addition, the outcomes showed that LMB restrained P release from the sediments, therefore, a portion of P-mobile is reduced in sediments and undergo transformation to non-reactive species [10]. According to Liu et al. [70], the effectiveness of LMB was tested and evaluated both in laboratory and field trials. The results of static and dynamic simulation experiments under different environmental conditions showed that with the application rate of LMB at 0.5 kg/m^2, the orthophosphate concentration of the overlying water decreased to a low level ($\leq$0.02 mg/L) within 10 days. Even under anaerobic and high pH (pH = 9.0) conditions, the phosphate release suppression efficiency reached 98.3%, and the P-release rate was −8.20 mg/m^2d (negative value indicates P adsorption by LMB). The monitoring data of the field sediments rehabilitation project were consistent with the results achieved in laboratory experiments, thus showing that the application of LMB could inhibit the internal P release effectively [70].

5.3.2 Field Studies

One of the first field applications of the LMB was undertaken in 2002 in two estuaries along the coastline of southwest Western Australia [71]. The LMB reduced the dissolved P in the water column to below the detection limit in a few hours, and the P flux from the sediments dropped substantially during the trial. The effect of phosphorus reduction on phytoplankton growth was clear in the phytoplankton-dominated Vasse River, but was less evident in the alternating phytoplankton to aquatic plant-dominated Canning River which is also subject to surface nutrient inputs [71].

The LMB treatment of the lake Het Groene Eiland (The Netherlands) (Table 5.4 at Appendix), which was created in winter 2008 after the construction of three dykes that isolated this area from the surrounding water body, had no or only a marginal effect on TP and SRP concentrations [72]. Mean TP and SRP levels in the treated and adjacent lake were identical. There were some confounding factors to address which include algal growth and overwintering waterfowls [72]. This case establishes that proper assessment must be taken before implementing reform. This suggests the investigation of the water and nutrient flows, the biological and biotic parameters of the system, and the ecosystem background for the particular water source [72].

Very high removal efficiency (80–95%) of TP and SRP within 2 weeks after LMB treatment was reported by Bishop et al. [73] for the Laguna Niguel Lake (Table 5.4 at Appendix), California, USA. Phoslock^{TM} was able to rapidly (<2 weeks) and significantly decrease total (>80%) and free reactive (>95%) phosphorus in the water column and shift potentially releasable sediment phosphorus fractions to residual forms after treatment (Fig. 5.10). Despite the cyanobacteria events and large suspension cells density before LMB application, cyanobacteria concentrations were low or almost undetectable after application [73].

In the relevant literature, the short-term alterations in elemental composition and P-fractions of the sediment in a field study in the Clatto Reservoir in Scotland (Table 5.4 at Appendix) was also investigated. Sediment core samples were pre-taken 2 days before and post-taken 28 days after 24 tonnes of LMB application at the 9 ha reservoir [74]. For the post-taken samples, a substantial increase of La content in the upper 8 cm of sediment was shown, thus, theoretically increase the P-binding capacity in sediment by 250 kg for the whole reservoir scale. Moreover, the theoretical estimation of mobile P binding (release-sensitive) by mass La balance calculation, yielded a potential binding of 42% mobile P being presented in the top 4 cm, or 17% mobile P being presented in the top 10 cm. While the fractions in the top 2 cm sustained a significant increase in residual P, the sediment P-fractions, involving that of mobile P, did not show noticeable fluctuation following the application of LMB. After conducting experimental conditions to P extraction from LMB, it was noticed that a portion of 21% P bound by modified clay could be easily released while the remaining 79% was impossible to be released at pH value ranges of 5–9 in shallow lakes. Therefore, LMB application are proved capable to increase the P sorption capacity of sediments under reducing conditions [74].

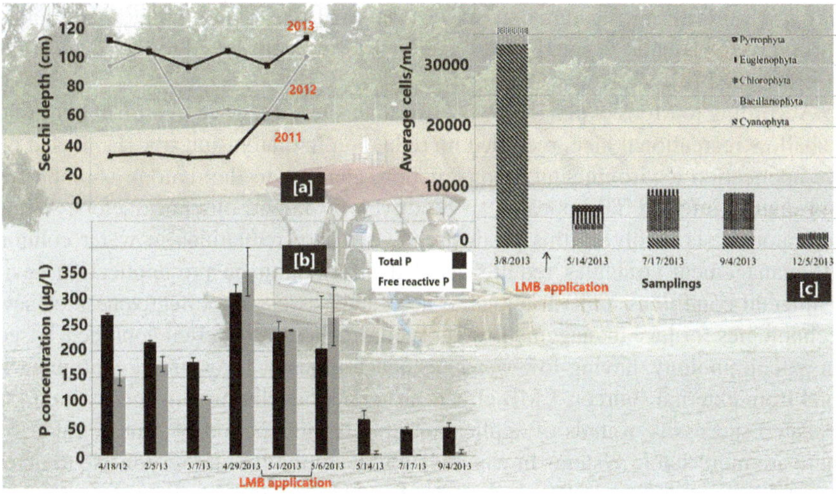

Fig. 5.10 **a** Secchi depth measurements over 3 years pre- and post-LMB application in Laguna Niguel Lake, **b** total and free reactive phosphorus concentrations throughout Laguna Niguel Lake, **c** algae assemblage composition at different sampling events in Laguna Niguel Lake. Modified from [73]

Another technique of LMB application, that of "Flock and Lock" was reported for the lake Rauwbraken, the Netherlands (Table 5.4 at Appendix). This technique combines a low dose of the polyaluminumchloride flocculent and the LMB [75]. The aforementioned water restoration technique involved the inactivation of the water column phosphate, the precipitation of particulate P, followed by the phosphate immobilization that is released from the precipitated flocks and the sediment. The technique is proven effective in stripping both dissolved (phosphate) and particulate P from the water column, but it is also countering the internal loading by fixing the P release from the sediment in a permanent way. Therefore, a complete P removal from the water column and an extensive P reduction release from the sediments have been reported. Besides, the simultaneous filamentous cyanobacteria removal and chlorophyll-a dropping to 2 μg/L are promising an improved water quality through the treatment proposed. However, temporal elevation of Al, La, and suspension of clay quantities were observed. Based on this technique, total elimination of grazing phytoplanktonic species (*Daphnia galeata*) from the water column one week after the application had been reported, and it had been kept three months afterward. Such a species disappearance from the lake are caused by jointly considering the physical effects due to flocks, grazing inhibition by flocks and clay, very low food quantities, as well as the absence of predation refuge. The rebounding to the pre-treatment conditions have been also literature reported [75]. The "Flock and Lock" was further applied in the Lake De Kuil, the Netherlands, causing a significant increase in macrophyte coverage from almost no macrophytes up to a covered area of 12% for the period of two years after the treatment. *Elodea nuttallii* and *Chara vulgaris* were the most dominant species reported over time, while a substantial portion of the aquatic vegetation of Lake De Kuil was attributed to the filamentous macro-algae [76].

According to Moos et al. [77], two Australian sites, a sewage treatment pond and a shallow recreational lake, recorded no reduction in diatom-inferred TP. Based on pre-application environmental reconstruction, changes to the diatom assemblages and diatom-inferred TP appeared to be driven by larger, climatic factors. While laboratory tests involving this product showed sharp reductions in water column TP, management strategies require detailed information on pre-application environmental conditions and variations to accurately assess the effectiveness of new technologies for lake management [77]. According to this different approach, based on paleolimnology, having low water residence times at lakes and a continuing P load from external sources, LMB effects are rapidly declining, thus necessitating a repeated successive rounds of application after a short period of time. Contrarily, there are more stable systems in which diatom communities are responsive to other than traditional drivers of P reduction, such as the variation of climatic conditions [77].

Epe et al. [56] developed a long-term study in examining the water quality characteristics of a polymictic, eutrophic swimming lake in the Lake Bärensee, central Germany, which is an artificially excavated lake that suffers from frequent cyanobacterial blooms. An initial application of LMB is dating back to 2007, where smaller repetitive applications of LMB were effected in 2010 and 2013, where frequent external nutrient loads caused excessive amounts of P, beyond defined levels. Moreover, there was an decreased intensity of the process of mineralization, thus, preventing large algal and cyanobacterial blooms. The post-treatment indicators of chlorophyll-a, total N, and ammonium implied the limited lake productivity by recurring short treatments with LMB. In parallel, the aforementioned effectiveness of P management enabled the safe and continuous utilization of the lake for swimming and recreation activities [56].

From the analysis above, it can be concluded that the applications of Phoslock™ to water sources are proven especially effective regarding nutrients' management, primary P-control and the concurring physicochemical conditions of pH, chlorophyll-a, Secchi disk depth, among others. It can be soundly argued that Phoslock™ can be proven a certainly promising material for lakes' restoration from the P-induced eutrophication.

5.3.3 Cost Evaluation of Phoslock™

A technical and economic evaluation of Phoslock™ can reveal those subtle parameters that are proven of utmost importance in achieving lake restoration [10]. Based on Table 5.3, the authors Lewtas et al. [78] and Zamparas et al. [10] have suggested that the "surface area" and "mean depth" variables are analogously contributing to Phoslock™ load and concentration, where the representative Phoslock™ weighted

Table 5.3 Phoslock™ load and concentration upon 15 literature-selected lakes at four countries worldwide. *Source* Modified from [10]

Country	No. of lakes	Weight: [Surface area (ha)] X [Mean depth (m)]	Phoslock™ load range (tons/ha)	Weighted Phoslock™ load (tons/ha)	Weighted Phoslock™ concentration (mg/L)
United Kingdom (UK)	3	9.9–25.2	1.6–3.0	40.6	1865.4
The Netherlands (NL)	2	28.0–35.2	4.5–5.9	108.1	1986.7
Germany (G)	9	2.1–150.8	1.9 – 6.4	303.5	5550.9
New Zealand (NZ)	1	680.0	0.06	0.06	0.29
Greece (GR), own study	1	1.1	–	3.7 (estimation)	–

loads and concentrations are counted of the corresponding weighted values. These weighted values are yielded by the combined calculation of [surface areas] and [mean depth] for 15 lakes, being allocated at the four countries: United Kingdom, the Netherlands, Germany, and New Zealand (Table 5.2). Regarding the study of Lewtas et al. [78] and their analysis on the artificial pond covering an area of 0.5 ha of mean depth 2.1 m, it was measured that in a mean depth lake of 2.4 m, the Phoslock™ load can support a strong efficiency of P-binding. Besides, at the range of 150–200 mg L^{-1} (mean value 163 mg L^{-1}), the application of 3.5–4.0 tons ha^{-1} (mean value 3.7 tons ha^{-1}) of Phoslock™ is needed, compared to the excessive use of natural materials N-Zeolite and N-Bentonite needed, without meeting such P-binding efficiency, as Phoslock™ achieves [10].

The weights stated above are relative weights to each other, having no absolute identity, thus, no normalization is needed (since the numerator and denominator coefficients are mutually canceled out). Moreover, it is noteworthy that the cost of Phoslock™ treatment depends on a plethora of parameters, such as the application dosage, the amount to be monitored, the biologically available and immobilized P, the data analysis, and the costs of labor and equipment, that are linked to lakes' size [78].

In the study of Lewtas et al. [78], it was argued that the calculation of Phoslock™ cap material per m^2 is related to dosage of 100 g Phoslock™/g mobile sediment P. Assuming that the quantity of mobile P in the top 5 cm ranges at 50–400 mg kg^{-1}, then, the cost ranges at $0.75–6.02 (showing a linear pricing behavior) respectively.

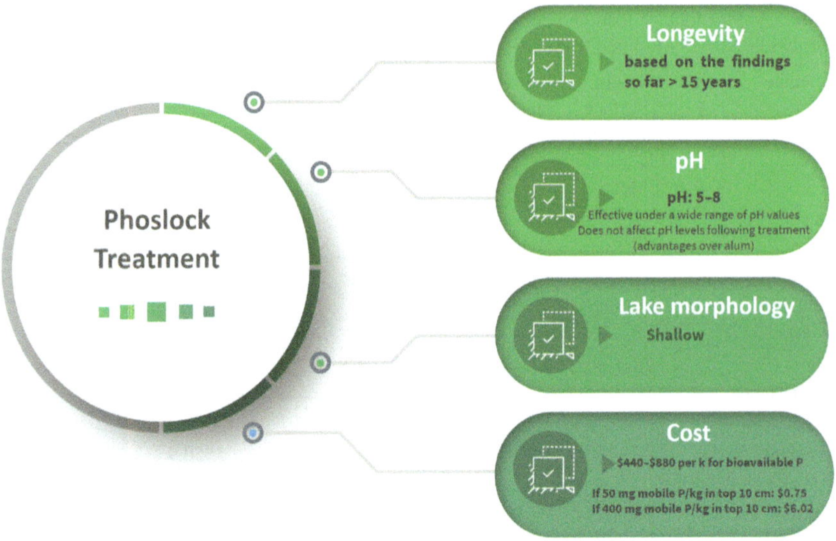

Fig. 5.11 Phoslock^{TM} treatment summary. *Source* Authors' own study

Therefore, according to Zamparas et al. [10], in the lake studied (of 0.5 ha surface area, top 10 cm depth, and 203.53 mg mobile P kg^{-1}), its dimension equals to 250 m^3 volume of sediment-layer, or 250,000 kg of sediment-layer, and 50.88 kg of mobile sediment P. After an econometric calculation, it was found that the quantity of Phoslock^{TM} applied was 5.088 tons and the price of the applied Phoslock^{TM} applied was \$30.53 kg^{-1} Phoslock^{TM} [78]. It was estimated that the quantity of the applied 5 tons Phoslock^{TM} is costing 155.34 thousand dollars, whereas Lewtas et al. [79] argued that the dosage of applied Phoslock^{TM} was 100 times higher than the quantity of mobile sediment P treated with a cost of \$7.5–60.2 (Fig. 5.11).

5.4 Other Materials for Lake Remediation

5.4.1 Ferric Chloride

Through precipitation with a combination of ferric salt and lime, ferric chloride has been successfully used to extract phosphates from sewage effluent. It has also been used in lake waters [79–81]. The reactions of ferric chloride with bicarbonate alkalinity and phosphate are presented in Table 5.1.

The stoichiometric equation (Table 5.1, Eq. I) is believed to predominate at pH above 7 and indicates that 1 kg of P requires 1.8 kg of ferric iron. However, the addition of ferric iron will also result in reduced lake water alkalinity with precipitation of ferric hydroxide (Eq. II), and the efficiency of water column phosphate removal may be low with both reactions occurring (Table 5.1). The addition of a suitable buffer may be required for low alkalinity lakes.

In a recent study by Li et al. [40], the control efficacy of phosphorus by ferric chloride in Dianchi Lake sediments was examined. The effects of the dosage of ferric chloride and contact time on P-release from sediments were studied. Laboratory trials showed that ferric chloride can successfully prevent the release of phosphorus from sediments. At a ferric chloride amount of 10 mg/g, the total phosphorus concentration of the overlying water was reduced by ~87%. In column experiments, by the addition of 6 mg/g of ferric chloride, the total phosphorus concentration of the overlying water was still less than 0.01 mg/L after 100 days. Moreover, sediment analysis revealed that ferric chloride mainly converts unstable exchangeable phosphorus (Ex–P), ferric phosphate (Fe–P), and organic phosphorus into more stable occluded phosphate (O–P), reducing the possible release of phosphorus from sediments [40].

The biogeochemical environment in which the absorbent may be deployed can have an effect on the efficiency of an active capping layer. The high content of Fe (III) oxyhydroxides in various iron-containing absorbents is a major factor in their efficacy. Under reducing environments, such as those present in anoxic hypolimnion sediments of seasonally stratified lakes, Fe (III) is microbially reduced to Fe (II), releasing adsorbed P. Thus, unless the procedure is coupled with adequate aeration to sustain oxic conditions in the hypolimnion throughout the long term (weeks to months), Fe (III) absorbents are likely to be unsuccessful. The addition of Fe to the anoxic hypolimnion, or sediments, has the added advantage of allowing the incredibly toxic H_2S to precipitate as insoluble FeS. This treatment can minimize the negative effects of turnover and sediment microbial processes on whole-lake waters. In addition, a recent study reported gradual long-term (20 years) improvements in water quality following a single bulk Fe addition, despite annual anoxic events, which diminished as a consequence of less organic matter accumulation [82]. However, the slowness and dependence of such gradual organic matter decrease will not always meet the ambitions of water managers for rapid outcomes.

5.4.2 Calcite

When photosynthesis drives pH to values above pH 9.0–9.5, calcite will precipitate with possible co-precipitation of P. The formed P precipitates, such as hydroxyapatite, have low solubility at pH higher than 9 and consequently, application of Calcite to manage P is primarily suited for hard water lakes with relatively high pH, lowering sediment P release substantially. Laboratory experiments with sediment from eutrophic German lakes showed that active calcite barriers could strongly reduce the P flux from the sediment [83]. Several calcite characteristics like grain size, specific surface, and structure determine the P adsorbing capacity, while complexation can be hindered by organic matter and iron-ions [83]. A combined sodium aluminate and Ca hydroxide treatment applied to the hypolimnion strongly reduced the P diffusion rates from the sediment of Tiefwarensee and caused a rapid trophic shift that lasted at least for 7 years [84].

5.4.3 *Bephos*^{TM} *and Modified Local Soil (MLS)*

Over the last few decades, a series of modified local soil (MLS) materials that can be used to remove HABs and remediate polluted sediment have been developed [85]. The MLS materials are made from clean local soil, or commercially available clay/solid particles, that are modified by a very small amount of natural products (usually less than 1% of the soil), such as chitosan, cationic starch, Moringa oleifera, and xanthan [86]. The solids can be local soils, local sands, clays, red soil, or coal fly ash [87, 88]. MLS is also used for capping so that the nutrient release from sediment to water can be reduced. After the MLS flocculation and capping treatment, the decomposition of algae toxins remarkably increases in the sediment compared to bulk water treatment. Subsequently, when the water clarity is achieved, the capped algal flocs and associated nutrients can be utilized in situ as fertilizers for the growth of submerged macrophytes in shallow water systems, which is beneficial as macrophytes can support a healthier food chain than algal blooms. Therefore, MLS technology has made it possible to put algae along with the excessive nutrients into the food chain by flocculating and removing them from water into sediment and converting them into submerged vegetation [89]. Recently, an interface oxygen nanobubble technology has been developed, which allows the loading of a substantial amount of oxygen into the microporous surfaces of clays [90]. These clays can stably hold the nanobubbles for months, so that once the oxygen is delivered by settling the clay particles onto the sediment, an aerobic capping layer is formed on top of the anaerobic sediment. This oxidized capping layer can prevent the consumption of dissolved oxygen in the water column by the anaerobic substances from the sediment underneath the capping layer [90]. Furthermore, it has been demonstrated that emissions of the greenhouse gases, CH_4 and CO_2, from eutrophic waters can be substantially reduced through the oxygen nanobubble-MLS treatment. Thus, the oxygen nanobubble-MLS treatment may provide a principle for delivering oxygen down into deep waters or sediment without the need for mixing energy (e.g., pumps) to combat anoxia/hypoxia problems that are crucial for eutrophication control [91].

5.5 Conclusions

Based on the results of the above long-term and detailed studies, we can conclude that one of the major challenges in determining the true effectiveness of initiatives to reduce P loading to lakes is a lack of pre- and post-treatment monitoring, especially in terms of nutrient bioavailability. In other words, since only certain types of phosphorus are biologically available for phytoplankton development, using total phosphorus concentrations to estimate eutrophication risk can be problematic for management. Although total phosphorus (TP) is widely calculated in surface waters and is the basis for most water quality requirements, it does not differentiate between different types of P, which can have substantially different bioavailability. Although total phosphorus (TP) is commonly calculated in surface waters and is the basis for most water quality requirements, it does not differentiate between different types of P, which can have substantially different bioavailability. When a comprehensive evaluation of sources is conducted using mass balance and seasonal accounting, TP can be an appropriate indicator. To better assess the multitude of factors that can affect the durability of existing nutrient reduction initiatives and enhance future restoration outcomes, adequate monitoring systems that track both external and treated flows and differentiate between bioavailable and recalcitrant nutrient forms are needed. Finally, cost plays an important role in deciding which restoration method to use to minimize a specific nutrient source, as well as which form of source control (internal vs. external) would produce the most cost-effective outcome.

Figure 5.12 summarizes the basic characteristics of material behavior in natural aquatic ecosystems. Thus, we can evaluate the functionality of each inactivation agent in terms of conditions under which it can be used effectively as a restoration tool in eutrophic waters. In fact, we focus on the defects in each material in order to choose the best method for implementation.

The reduction of DRP concentration in the water, the period of P-inactivation and immobilization, potential adverse effects on the phytoplankton population, and possible toxicity from high Al^{+3} concentrations can all be used to determine the overall efficacy of an alum application.

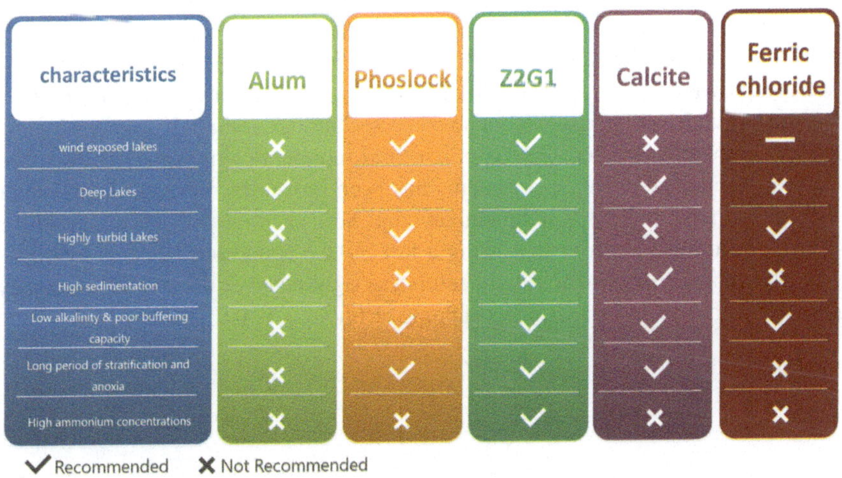

Fig. 5.12 Basic features of P-inactivation agents in natural water bodies, evaluating the effectiveness in relation to morphological and physicochemical characteristics of a eutrophic ecosystem. *Source* Modified from [19]

The material's key benefits are its low cost and the fact that, in addition to lake surface applications, it can also be added directly to the hypolimnion as a slurry, reducing the lake benthos' exposure to the floc.

The toxicity of alum, the decrease of lake water pH, which necessitates buffer dosing, poor floc stability due to wind mixing, and harmful effects on the littoral benthos from epilimnion treatment are just a few of the drawbacks. Moreover, smothering of the benthos by the settling floc, potential resuspension of the floc in shallow lakes, restriction period for drinking, irrigation, and stock water use after an application are also serious drawbacks. The study of 114 alum-treated lakes showed that treatment longevity was slightly longer in deeper, stratified lakes (mean 21 years) than in shallow, polymictic lakes (mean 5.7 years), where sediment disturbance by bottom-feeding fishes (carp) decreased treatment longevity.

Phoslock^{TM} has high efficiency in removing SRP and a high potential for sediment capping, preventing P release. The field applications showed positive results in the purification of eutrophic water. With Phoslock^{TM}, eutrophic water could be transformed to good water quality, allowing for the growth of several aquatic plants.

Moreover, Phoslock^{TM} could eventually purify the eutrophic water, thus, it was an effective way for water eutrophication prevention. Also, in many recent studies, most La was recovered in the HCl extract, indicating that P became strongly bound to La in the Phoslock^{TM} matrix [92, 93]. A detailed evaluation of Phoslock^{TM} treated lakes found reductions in surface water TP (data available from 15 lakes), SRP (in 14 lakes), chlorophyll-a concentrations (in 15 lakes) with increased Secchi disk depth (in 15 lakes). Furthermore, increase in aquatic macrophyte community species numbers (average increase of 1.6 species; in 6 lakes) and maximum colonization depths (mean increase of 0.7 m; in 4 lakes) were reported.

As far as the efficacy and the terms of sediment capping are concerned, we have a fast reduction of phosphate concentration in the sediment pore water present inside the calcite active barrier. This is because the higher the P concentration, the faster is the precipitation of calcium phosphate. Hence the rich in P sediment–water interface is highly recommended for ABS application. Furthermore, the use of an ABS particularly with calcite, a relatively novel ecotechnology, definitely is an ecological application mainly because $CaCO_3$ is a natural sediment component [83, 94]. The survival of macrozoobenthic organisms has been proven by Berg et al. [83] after applying a calcite ABS for sediment capping. Still, there is a disadvantage in the stability of an active barrier that being a few in-lake conditions, such as hydrodynamics and bioturbation. These two factors can significantly diminish the efficiency and sustainability of this sediments capping technology [94].

Modified zeolite (Z2G1) and Phoslock^{TM} settle rapidly, and this is an advantage in highly turbid lakes where sediment capping is the main priority. For example, Phoslock^{TM} increased the sediment stability threshold by 265% whereas Al had no stabilizing effect. However, it could be a disadvantage in lakes with high inputs as the Z2G1 and Phoslock^{TM} capping layer would become buried (Fig. 5.12). In addition, more rapid settling may allow less time for binding phosphate as it sinks into the water column [95].

Recently, there have been renewed arguments to suggest that P removal alone may not alleviate eutrophication and to recognize the importance of simultaneous control of P and N [96]. It has been observed that controlling only phosphorous (P) inputs to freshwaters and ignoring the large anthropogenic inputs of Nitrogen (N) can reduce algal uptake of N and thus, allow more N to be transported downstream where it can exacerbate eutrophication problems in estuarine and coastal marine ecosystems [97]. Based on this fact, improvements in the water quality of many freshwater and most coastal marine ecosystems require a decrease in both nitrogen and phosphorus inputs. Thus, the capacity of modified zeolite for adsorption of both phosphorus and ammonia is an important advantage against other materials used to restore eutrophic water bodies. Moreover, in cases where one of the two factors, e.g., P or N, is the limiting nutrient of a natural water ecosystem, the use of Z2G1 is appropriate without requiring additional use of other materials.

Appendix

See Table 5.4.

Table 5.4 P-inactivation materials and implementation actions to natural eutrophic waterbodies. Indicative field applications with Phoslock^{TM}, modified zeolite Z2G1, Alum, and Calcite. Source: modified from [19]

Lake/reservoir	Application year	Morphological parameters	Application (tn)	Material	Key features	References
Swan Canning and Vasse Wonnerup estuaries, Australia	2001	MD: 2 m depth	45 and 40 respectively	Phoslock^{TM}	Use of Phoslock^{TM} in removing dissolved phosphorus (P) from the water column as well as to intercept sediment, explores the effect of the reduction of P availability on phytoplankton activity	[56]
Lake Dianchi, China	2006	0.2 ha 11 m Max 5 m mean	10	Phoslock^{TM}	Phoslock^{TM} has great efficiency in SP removal, normally 95%, and effective capacity to sediment capping of the bottom slurry for avoiding the nutrient release. With the treatment by Phoslock^{TM}, the eutrophic water could be transferred to good water quality with variously healthy aquatic plants growing	[98]
Cane Parkway pond, Canada	2008	0.4 ha 2 m mean	–	Phoslock^{TM}		[79]
Uki, Australia	2008	1.4 ha 1 m mean	4			[79]

(continued)

Table 5.4 (continued)

Lake/reservoir	Application year	Morphological parameters	Application (tn)	Material	Key features	References
De Ploeg urban pond, The Netherlands	2009	0.16 ha MD: 1 m depth		Phoslock^{TM}	Enclosure experiment was conducted to test the effectiveness of dredging and a Phoslock^{TM} application in controlling cyanobacterial blooms and toxin production	[66]
Clatto Reservoir Dundee, UK	2009	9 ha MD: 2.75 m depth	24	Phoslock^{TM}	Short term alterations in sediment La composition and sediment P-fractions before and after the application of Phoslock^{TM}	[76]
Rauwbraken, The Netherlands	2008	2.6 ha 8 m depth		Phoslock^{TM}	A novel restoration technique "Flock & Lock" was evaluated	[99]
Loch Flemington, Scotland	2010	15 ha mean depth 0.75 m and maximum depth of 2.35 m.	25	Phoslock^{TM}	During summer 2010, the TP concentration decreased by 65% and in 2011 by 81% from the summer 2009 values (prior to the LMB application). Chlorophyll-a was also significantly reduced by 77% and the following summer there was a 95% reduction from levels recorded in 2009.	[100]

(continued)

Table 5.4 (continued)

Lake/reservoir	Application year	Morphological parameters	Application (tn)	Material	Key features	References
Laguna Niguel Lake	2013	1.2 ha 9.5 m max 3.7 m mean	51.34	Phoslock™	Laguna Niguel Lake allowed evaluating the operational effectiveness of LMB in a system historically plagued by high phosphorus concentrations, potentially toxic cyanobacteria (Aphanizomenonflos-aquae dominant)	[75]
Lake Okaro, Central North Island, New Zealand	2011	32 ha	110 (equivalent to a dose rate of 350 g m^{-2})	Modified zeolite (Z2G1)	Measuring sub-lethal effects on freshwater crayfish (*Paranephrops planifrons*) behavior and physiology of the exposure to modified zeolite	[101]
Lake Okaro, Central North Island, New Zealand	2011	32 ha	110 (equivalent to a dose rate of 350 g m^{-2})	Modified zeolite (Z2G1)	Effects of a modified zeolite Z2G1on P and N processes and fluxes across the lake sediment-water interface	[102]
Lake Okaro, Central North Island, New Zealand	2011	32 ha	110 (equivalent to a dose rate of 350 g m^{-2})	Modified zeolite (Z2G1)	Effects of sediment capping on zooplankton and phytoplankton	[102, 103]

(continued)

Table 5.4 (continued)

Lake/reservoir	Application year	Morphological parameters	Application (tn)	Material	Key features	References
Lake Okaro, Central North Island, New Zealand	2011	32 ha	110 (equivalent to a dose rate of 350 g m^{-2})	Modified zeolite (Z2G1)	Measuring the in situ health of two finfish species (Oncorhynchus mykiss and Gobiomorphus cotidianus) and crayfish (Paranephrops planifrons) with respect to a large-scale application of Z2G1	[104]
Lake Okaro, Central North Island, New Zealand	2008	32 ha	110 (equivalent to a dose rate of 350 g m^{-2})	Alum	Impact of the material on the water nutrient concentrations and on the trophic status indicator Chlorophyll-a	[105]
Lake Süsser See, Germany	2003	268 ha 8.2 m max 4.3 m mean	Alum treatments were conducted annually or biannually from 1977 to 1992, with 500–700 t precipitant (2 mg Al L^{-1} lake volume) added in spring, a total of 7000 t	Alum	Long term effects of phosphorus precipitations with Alum in hypereutrophic lake	[33]
Newman Lake, Washington State, USA	2009	50 ha	~200 t/year	Alum	Alum introduction into an oligomictic lake. Whole lake application and microfloc installation	[36, 37]

(continued)

Table 5.4 (continued)

Lake/reservoir	Application year	Morphological parameters	Application (tn)	Material	Key features	References
Green Lake Washington State, USA	2008	105 ha 8.2 m max 3.9 m mean	94 t	Alum	Short-term effects of buffered alum treatment on lake sediment phosphorus speciation	[39]
Lake Luzin, Germany	2011	58 ha 33 m max 18.1 m mean	300 t of $Ca(OH)_2$ were added to thehypolimnio	Calcite	Dynamic of phosphorus and its binding forms in seston and sediment before and during the whole lake treatment	[106]
Lake Arendsee (Germany)	2000	513 ha 49 m max 29 m mean	25,000 tons of sediments from calcite rich deposits (Seekreide) from near the northern shore was applied to the surface of the lake with a water depth below 25 ± 30 m (>65% of lake area)	Calcite	A failed restoration technique where calcareous mud from the littoral zone was resuspended and distributed over the entire lake surface	[107]
North Halfmoon Lake and Lofty Lake (Alberta, Canada)	1998	80 ha 6.6 m max 3.1 m mean	197 t	Calcite	The relationships between zooplankton and phytoplankton, together with the effects of lime additions during a 3-year study (1991–93)	[108]

(continued)

Table 5.4 (continued)

Lake/reservoir	Application year	Morphological parameters	Application (tn)	Material	Key features	References
Eight hardwater lakes monitored in this study: (1) Crooked Lake, (2) Jenkins Lake, (3) Baptiste Lake, (4) Nakamun Lake, (5) Lofty Lake, (6) N. Halfmoon Lake, (7) Halfmoon Lake, and (8) Figure Eight Lake	2001	(7) 8.5 m max, (8) 6m max		Calcite	Effects of single and multiple moderate doses of lime (slaked lime, $Ca(OH)_2$, and/or calcite, $CaCO_3$) on eutrophic hardwater lakes	[109]

References

1. Dang C, Lu M, Mu Z, Li Y, Chen C, Zhao F, Yan L, Cheng Y (2019) Phosphorus fractions in the sediments of Yuecheng Reservoir, China. Water (Switzerland)
2. Cao X, Song C, Xiao J, Zhou, Y (2018) The optimal width and mechanism of riparian buffers for storm water nutrient removal in the Chinese eutrophic Lake Chaohu watershed. Water (Switzerland)
3. de Sosa LL, Glanville HC, Marshall MR, Abood SA, Williams AP, Jones DL (2018) Delineating and mapping riparian areas for ecosystem service assessment. Ecohydrology
4. Zhong JC, Yu JH, Zheng XL, Wen SL, Liu DH, Fan CX (2018) Effects of dredging season on sediment properties and nutrient fluxes across the sediment-water interface in Meiliang Bay of Lake Taihu, China. Water (Switzerland)
5. Preece EP, Moore BC, Skinner MM, Child A, Dent S (2019) A review of the biological and chemical effects of hypolimnetic oxygenation. Lake Reserv Manag
6. Xiong C, Wang D, Tam NF, Dai Y, Zhang X, Tang X, Yang Y (2018) Enhancement of active thin-layer capping with natural zeolite to simultaneously inhibit nutrient and heavy metal release from sediments. Ecol Eng
7. Lan C, Chen J, Wang J, Guo J, Yu J, Yu P, Yang H, Liu Y (2017) Application of circular bubble plume diffusers to restore water quality in a sub-deep reservoir. Int J Environ Res Public Health
8. Moore BC, Cross BK, Beutel M, Dent S, Preece E, Swanson M (2012) Newman Lake restoration: A case study part III. Hypolimnetic oxygenation. Lake Reserv Manag
9. Mobley M, Gantzer P, Benskin P, Hannoun I, McMahon S, Austin D, Scharf R (2019) Hypolimnetic oxygenation of water supply reservoirs using bubble plume diffusers. Lake Reserv Manag
10. Zamparas M, Kyriakopoulos GL, Kapsalis VC, Drosos M, Kalavrouziotis IK (2019) Application of novel composite materials as sediment capping agents: column experiments and modelling. Desalin Water Treat
11. van Oosterhout F, Waajen G, Yasseri S, Manzi Marinho M, Pessoa Noyma N, Mucci M, Douglas G, Lürling M (2020) Lanthanum in water, sediment, macrophytes and chironomid larvae following application of Lanthanum modified bentonite to lake Rauwbraken (The Netherlands). Sci Total Environ
12. Meis S, Spears BM, Maberly SC, Perkins RG (2013) Assessing the mode of action of Phoslock® in the control of phosphorus release from the bed sediments in a shallow lake (Loch Flemington, UK). Water Res
13. Cooke GD, Welch EB, Peterson S, Nichols SA (2016) Restoration and management of lakes and reservoirs
14. Jeppesen E, Søndergaard M, Jensen JP, Havens KE, Anneville O, Carvalho L, Coveney MF, Deneke R, Dokulil MT, Foy B et al (2005) Lake responses to reduced nutrient loading—an analysis of contemporary long-term data from 35 case studies. Freshw Biol
15. Cooke DG, Welch EB, Peterson SA, Nichols SA (2005) Restoration and management of lakes and reservoirs, 3rd edn. CRC Press, Boca Raton. ISBN 0-7484-0772-3
16. Søndergaard M, Jensen JP, Jeppesen E (2003) Role of sediment and internal loading of phosphorus in shallow lakes. In: Proceedings of the hydrobiologia
17. Huser BJ, Egemose S, Harper H, Hupfer M, Jensen H, Pilgrim KM, Reitzel K, Rydin E, Futter M (2016) Longevity and effectiveness of aluminum addition to reduce sediment phosphorus release and restore lake water quality. Water Res
18. Hickey CW, Gibbs MM (2009) Lake sediment phosphorus release management-Decision support and risk assessment framework. N Z J Mar Freshw Res
19. Zamparas M, Zacharias I (2014) Restoration of eutrophic freshwater by managing internal nutrient loads. A review. Sci Total Environ 496:551–562
20. Pilgrim KM, Huser BJ, Brezonik PL (2007) A method for comparative evaluation of whole-lake and inflow alum treatment. Water Res

21. Rydin E, Welch EB (1998) Aluminum dose required to inactivate phosphate in lake sediments. Water Res
22. Sanz J, Lombraña JI, De Luis AM, Ortueta M, Varona F (2003) Microwave and Fenton's reagent oxidation of wastewater. Environ Chem Lett
23. Bouyer F, Sanson N, Destarac M, Gérardin C (2006) Hydrophilic block copolymer-directed growth of lanthanum hydroxide nanoparticles. New J Chem
24. Pitre D, Boullemant A, Fortin C (2014) Uptake and sorption of aluminium and fluoride by four green algal species. Chem Cent J
25. Han J, Jeon B, Futatsugi N, Park HD (2013) The effect of alum coagulation for in-lake treatment of toxic Microcystis and other cyanobacteria related organisms in microcosm experiments. Ecotoxicol Environ Saf
26. Soumis-Dugas G, Triffaut-Bouchet G, Galvez R, Martel L (2009) Ecotoxicological assessment of an in-lake remediation method. J ASTM Int
27. Narf RP (1990) Interactions of chironomidae and chaoboridae (diptera) with aluminum sulfate treated lake sediments. Lake Reserv Manag
28. Pilgrim KM, Brezonik PL (2005) Evaluation of the potential adverse effects of lake inflow treatment with alum. Lake Reserv Manag
29. Nogaro G, Burgin AJ, Schoepfer VA, Konkler MJ, Bowman KL, Hammerschmidt CR (2013) Aluminum sulfate (alum) application interactions with coupled metal and nutrient cycling in a hypereutrophic lake ecosystem. Environ Pollut
30. Galvez-Cloutier R, Saminathan SKM, Boillot C, Triffaut-Bouchet G, Bourget A, Soumis-Dugas G (2012) An evaluation of several in-lake restoration techniques to improve the water quality problem (eutrophication) of Saint-Augustin Lake, Quebec, Canada. Environ Manag
31. Gibbs MM, Hickey CW, Özkundakci D (2010) Sustainability assessment and comparison of efficacy of four P-inactivation agents for managing internal phosphorus loads in lakes: sediment incubations. Hydrobiologia 658:253–275
32. Welch EB, Cooke GD (1999) Effectiveness and longevity of phosphorus inactivation with alum. Lake Reserv Manag 15:5–27
33. Lewandowski J, Schauser I, Hupfer M (2003) Long term effects of phosphorus precipitations with alum in hypereutrophic Lake Süsser See (Germany). Water Res
34. Mires JM, Soltero RA, Keizur GR (1981) Changes in the zooplankton community of Medical Lake, WA, subsequent to its restoration by a whole-lake alum treatment and the establishment of a trout fishery. J Freshw Ecol
35. Smeltzer E (1990) A successful alum/aluminate treatment of Lake Morey, Vermont. Lake Reserv Manag
36. Moore BC, Christensen D (2009) Newman Lake restoration: a case study. Part I. Chemical and biological responses to phosphorus control. Lake Reserv Manag
37. Moore BC, Christensen D, Richter AC (2009) Newman Lake restoration: a case study. Part II. Microfloc alum injection. Lake Reserv Manag
38. Paul WJ, Hamilton DP, Gibbs MM (2008) Low-dose alum application trialled as a management tool for internal nutrient loads in Lake Okaro, New Zealand. N Z J Mar Freshw Res
39. Dugopolski RA, Rydin E, Brett MT (2008) Short-term effects of a buffered alum treatment on green lake sediment phosphorus speciation. Lake Reserv Manag
40. Li S, Lin Z, Liu M, Jiang F, Chen J, Yang X, Wang S (2020) Effect of ferric chloride on phosphorus immobilization and speciation in Dianchi Lake sediments. Ecotoxicol Environ Saf
41. Steinman AD, Ogdahl M (2008) Ecological effects after an alum treatment in Spring Lake, Michigan. J Environ Qual
42. Huser BJ, Futter M, Lee JT, Perniel M (2016) In-lake measures for phosphorus control: the most feasible and cost-effective solution for long-term management of water quality in urban lakes. Water Res
43. Brattebo SK, Welch EB, Gibbons HL, Burghdoff MK, Williams GN, Oden JL (2017) Effectiveness of alum in a hypereutrophic lake with substantial external loading. Lake Reserv Manag

44. Schütz J, Rydin E, Huser BJ (2017) A newly developed injection method for aluminum treatment in eutrophic lakes: effects on water quality and phosphorus binding efficiency. Lake Reserv Manag
45. Huser BJ (2017) Aluminum application to restore water quality in eutrophic lakes: maximizing binding efficiency between aluminum and phosphorus. Lake Reserv Manag
46. Cohen JM (1972) Nutrient removal from wastewater by physical chemical processes. Wiley Interscience
47. Valsami-Jones E (2015) Phosphorus in environmental technology: principles and applications. Water Intelligence Online, vol 4. ISBN 9781780402758
48. Melnyk PB, Norman JD, Wasserlauf M (1974) Lanthanum precipitation: an alternative method for removing phosphates from wastewater
49. Firsching FH, Kell JC (1993) The solubility of the rare-earth-metal phosphates in sea water. J Chem Eng Data
50. Diatloff E, Asher CJ, Smith FW (1993) Use of GEOCHEM-PC to predict rare earth element (REE) species in nutrient solutions. Plant Soil 155–156:251–254
51. Persy VP, Behets GJ, Bervoets AR, De Broe ME, D'Haese PC (2006) Lanthanum: a safe phosphate binder. Semin Dial
52. Wang Z, Shen D, Shen F, Li T (2016) Phosphate adsorption on lanthanum loaded biochar. Chemosphere
53. Goscianska J, Ptaszkowska M, Pietrzak R (2015) Equilibrium and kinetic studies of chromotrope 2R adsorption onto ordered mesoporous carbons modified with lanthanum. Chem Eng J
54. Fiol N, Villaescusa I (2009) Determination of sorbent point zero charge: usefulness in sorption studies. Environ Chem Lett
55. Robb M, Greenop B, Goss Z, Douglas G, Adeney J (2003) Application of Phoslock™, an innovative phosphorus binding clay, to two Western Australian waterways: preliminary findings. In: Proceedings of the hydrobiologia
56. Epe TS, Finsterle K, Yasseri S (2017) Nine years of phosphorus management with lanthanum modified bentonite (Phoslock) in a eutrophic, shallow swimming lake in Germany. Lake Reserv Manag
57. Zamparas M, Gianni A, Stathi P, Deligiannakis Y, Zacharias I (2012) Removal of phosphate from natural waters using innovative modified bentonites. Appl Clay Sci 62–63:101–106
58. Yamada-Ferraz TM, Sueitt APE, Oliveira AF, Botta CMR, Fadini PS, Nascimento MRL, Faria BM, Mozeto AA (2015) Assessment of Phoslock® application in a tropical eutrophic reservoir: an integrated evaluation from laboratory to field experiments. Environ Technol Innov
59. Spears BM, Mackay EB, Yasseri S, Gunn IDM, Waters KE, Andrews C, Cole S, De Ville M, Kelly A, Meis S et al (2016) A meta-analysis of water quality and aquatic macrophyte responses in 18 lakes treated with lanthanum modified bentonite (Phoslock®). Water Res
60. Copetti D, Finsterle K, Marziali L, Stefani F, Tartari G, Douglas G, Reitzel K, Spears BM, Winfield IJ, Crosa G et al (2015) Eutrophication management in surface waters using lanthanum modified bentonite: a review. Water Res
61. Behets GJ, Mubiana KV, Lamberts L, Finsterle K, Traill N, Blust R, D'Haese PC (2020) Use of lanthanum for water treatment A matter of concern? Chemosphere
62. Lürling M, Tolman Y (2010) Effects of lanthanum and lanthanum-modified clay on growth, survival and reproduction of Daphnia magna. Water Res 44:309–319
63. Álvarez-Manzaneda I, Baun A, Cruz-Pizarro L, de Vicente I (2019) Ecotoxicity screening of novel phosphorus adsorbents used for lake restoration. Chemosphere
64. Akhurst D, Jones GB, McConchie DM (2004) The application of sediment capping agents on phosphorus speciation and mobility in a sub-tropical dunal lake. Mar Freshw Res 55:715
65. Lürling M, Waajen G, Engels B, van Oosterhout F (2017) Effects of dredging and lanthanum-modified clay on water quality variables in an enclosure study in a hypertrophic pond. Water (Switzerland)

66. Vopel K, Gibbs M, Hickey CW, Quinn J (2008) Modification of sediment–water solute exchange by sediment-capping materials: effects on O_2 and pH. Mar Freshw Res 59:1101
67. Ross G, Haghseresht F, Cloete TE (2008) The effect of pH and anoxia on the performance of Phoslock®, a phosphorus binding clay. Harmful Algae
68. Dibtseva NM, Kienskaya KI, Nazarov VV (2001) Synthesis and some properties of sols prepared by hydrolysis of lanthanum nitrate. Colloid J
69. Reitzel K, Balslev KA, Jensen HS (2017) The influence of lake water alkalinity and humic substances on particle dispersion and lanthanum desorption from a lanthanum modified bentonite. Water Res
70. Liu B, Liu X, Yang J, Garman DEJ, Zhang K, Zhang H (2012) Research and application of in-situ control technology for sediment rehabilitation in eutrophic water bodies. Water Sci Technol
71. Robb M, Greenop B, Goss Z, Douglas G, Adeney J (2003) Application of Phoslock™, an innovative phosphorus binding clay, to two Western Australian waterways: preliminary findings. Hydrobiologia 494:237–243
72. Lürling M, van Oosterhout F (2013) Case study on the efficacy of a lanthanum-enriched clay (Phoslock®) in controlling eutrophication in Lake Het Groene Eiland (The Netherlands). Hydrobiologia
73. Bishop WM, McNabb T, Cormican I, Willis BE, Hyde S (2014) Operational evaluation of Phoslock phosphorus locking technology in Laguna Niguel Lake, California. Water Air Soil Pollut
74. Meis S, Spears BM, Maberly SC, O'Malley MB, Perkins RG (2012) Sediment amendment with Phoslock® in Clatto Reservoir (Dundee, UK): investigating changes in sediment elemental composition and phosphorus fractionation. J Environ Manag
75. Van Oosterhout F, Lürling M (2011) Effects of the novel "Flock & Lock" lake restoration technique on Daphnia in Lake Rauwbraken (The Netherlands). J Plankton Res
76. Waajen G, van Oosterhout F, Douglas G, Lürling M (2016) Geo-engineering experiments in two urban ponds to control eutrophication. Water Res
77. Moos MT, Taffs KH, Longstaff BJ, Ginn BK (2014) Establishing ecological reference conditions and tracking post-application effectiveness of lanthanum-saturated bentonite clay (Phoslock®) for reducing phosphorus in aquatic systems: an applied paleolimnological approach. J Environ Manag
78. Lewtas K, Roy D, Paterson M (2016) Manitoba Prairie Lakes: eutrophication and in-lake remediation treatments literature review. Int Inst Sustain Dev
79. Sherwood LJ, Qualls RG (2001) Stability of phosphorus within a wetland soil following ferric chloride treatment to control eutrophication. Environ Sci Technol
80. Wang Z, Li D, Qin H, Li Y (2012) An integrated method for removal of harmful cyanobacterial blooms in eutrophic lakes. Environ Pollut
81. Immers AK, van der Sande MT, van der Zande RM, Geurts JJM, van Donk E, Bakker ES (2013) Iron addition as a shallow lake restoration measure: Impacts on charophyte growth. Hydrobiologia
82. Kleeberg A, Herzog C, Hupfer M (2013) Redox sensitivity of iron in phosphorus binding does not impede lake restoration. Water Res
83. Berg U, Neumann T, Donnert D, Nüesch R, Stüben D (2004) Sediment capping in eutrophic lakes—efficiency of undisturbed calcite barriers to immobilize phosphorus. Appl Geochem
84. Wauer G, Gonsiorczyk T, Hupfer M, Koschel R (2009) Phosphorus balance of Lake Tiefwarensee during and after restoration by hypolimnetic treatment with aluminum and calcium salts. Lake Reserv Manag
85. Pan G, Miao X, Bi L, Zhang H, Wang L, Wang L, Wang Z, Chen J, Ali J, Pan M et al (2019) Modified Local Soil (MLS) technology for harmful algal bloom control, sediment remediation, and ecological restoration. Water (Switzerland)
86. Pan G, Zhang MM, Chen H, Zou H, Yan H (2006) Removal of cyanobacterial blooms in Taihu Lake using local soils. I. Equilibrium and kinetic screening on the flocculation of Microcystis aeruginosa using commercially available clays and minerals. Environ Pollut

87. Yuan Y, Zhang H, Pan G (2016) Flocculation of cyanobacterial cells using coal fly ash modified chitosan. Water Res
88. Dai L, Pan G (2014) The effects of red soil in removing phosphorus from water column and reducing phosphorus release from sediment in Lake Taihu. Water Sci Technol
89. Zhang H, Shang Y, Lyu T, Chen J, Pan G (2018) Switching harmful algal blooms to submerged macrophytes in shallow waters using geo-engineering methods: evidence from a ^{15}N tracing study. Environ Sci Technol
90. Shi W, Pan G, Chen Q, Song L, Zhu L, Ji X (2018) Hypoxia remediation and methane emission manipulation using surface oxygen nanobubbles. Environ Sci Technol
91. Wang L, Miao X, Ali J, Lyu T, Pan G (2018) Quantification of oxygen nanobubbles in particulate matters and potential applications in remediation of anaerobic environment. ACS Omega
92. Wang C, He R, Wu Y, Lürling M, Cai H, Jiang HL, Liu X (2017) Bioavailable phosphorus (P) reduction is less than mobile P immobilization in lake sediment for eutrophication control by inactivating agents. Water Res
93. Song X, Li D, Zhao Z, Zhou J, Xu C, Geng X, Huang Y (2020) The effect of microenvironment in the sediment on phosphorus immobilization under capping with ACPM and Phoslock®. Environ Sci Pollut Res
94. Yu X, Grace MR, Sun G, Zou Y (2018) Application of ferrihydrite and calcite as composite sediment capping materials in a eutrophic lake. J Soils Sediments
95. Özkundakci D, Hamilton DP, Gibbs MM (2011) Hypolimnetic phosphorus and nitrogen dynamics in a small, eutrophic lake with a seasonally anoxic hypolimnion. Hydrobiologia
96. Dodds WK, Smith VH (2016) Nitrogen, phosphorus, and eutrophication in streams. Inland Waters
97. Conley DJ, Paerl HW, Howarth RW, Boesch DF, Seitzinger SP, Havens KE, Lancelot C, Likens GE (2009) Ecology—controlling eutrophication: nitrogen and phosphorus. Science (80-)
98. Liu Y, Tian K, Winks A (2009) Water eutrophication prevention by phoslock flocculation and sediment capping reaction. In Proceedings of the 2009 international conference on environmental science and information application technology, ESIAT 2009
99. Lürling M, Oosterhout F (2013) Van Controlling eutrophication by combined bloom precipitation and sediment phosphorus inactivation. Water Res
100. Gunn IDM, Meis S, Maberly SC, Spears BM (2014) Assessing the responses of aquatic macrophytes to the application of a lanthanum modified bentonite clay, at Loch Flemington, Scotland, UK. Hydrobiologia
101. Parkyn SM, Hickey CW, Clearwater SJ (2011) Measuring sub-lethal effects on freshwater crayfish (Paranephrops planifrons) behaviour and physiology: laboratory and in situ exposure to modified zeolite. Hydrobiologia 661:37–53. https://doi.org/10.1007/s10750-010-0241-8
102. Gibbs M, Özkundakci D (2011) Effects of a modified zeolite on P and N processes and fluxes across the lake sediment–water interface using core incubations. Hydrobiologia 661:21–35. https://doi.org/10.1007/s10750-009-0071-8
103. Özkundakci D, Duggan IC, Hamilton DP (2011) Does sediment capping have post-application effects on zooplankton and phytoplankton? Hydrobiologia 661:55–64. https://doi.org/10.1007/s10750-009-9938-y
104. Landman MJ, Ling N (2011) Fish health changes in Lake Okaro, New Zealand: effects of nutrient remediation, season or eutrophication? Hydrobiologia 661:65–79. https://doi.org/10.1007/s10750-010-0315-7
105. Paul WJ, Hamilton DP, Gibbs M (2008) Low-dose alum application trialed as a management tool for internal nutrient loads in Lake Okaro, New Zealand. N Z J Mar Freshw 42:207–217
106. Dittrich M, Gabriel O, Rutzen C, Koschel R (2011) Lake restoration by hypolimnetic $Ca(OH)_2$ treatment: impact on phosphorus sedimentation and release from sediment. Sci Total Environ 409:1504–1515
107. Hupfer M, Pöthig R, Brüggemann R, Geller W (2000) Mechanical resuspension of autochthonous calcite (Seekreide) failed to control internal phosphorus cycle in a eutrophic lake. Water Res 34:859–867

108. Ghadouani A, Alloul BP, Zhang YP (1998) Relationships between zooplankton community structure and phytoplankton in two lime-treated eutrophic hardwater lakes. Freshw Biol 39:775–790
109. Prepas EE, Babin J, Murphy TP, Chambers PA, Sandland GJ, Ghadouani A et al (2001) Long-term effects of successive $CA(OH)_2$ and $CACO_3$ treatments on the water quality of two eutrophic hardwater lakes. Freshw Biol 46:1089–1103

Chapter 6
Non-invasive Removal of Phosphorus from Lakes Using Processed Calcite-Based Materials

Dorota Pryputniewicz-Flis, Agnieszka Bańkowska-Sobczak, Dorota Burska, Jakub Idźkowski, Łukasz Kozłowicz, and Grzegorz Brenk

Abstract In this study, milled ground calcium carbonate (GCC) consisting of calcite is developed as an orthophosphate (OP) binding agent to be used in a strategy for non-invasive orthophosphate (OP) removal from eutrophic lakes. Planetary ball mill and 74 different grinding methods were applied differing in grinding time (0.5–24 h), grinding force (150–400 rpm) and mass of grinding balls (150–500 g). The obtained materials had specific diameter d_{90} of 7.3–86.6 μm and specific surface area (SSA) of 4.9–24.0 m^2/g. Grinding resulted in 98% OP removal, being five times higher compared to the initial GCC material (21%). Grinding substantially improved OP removal's ability as compared to the source GCC, without using chemicals, and at low cost. Such GCC materials were used to develop a strategy for non-invasive OP removal from the aquatic environment which assumes application of a GCC containing carrier into the water body and its subsequent removal. Three different solutions were analysed, differing in type and the adjustment of the carrier's adhering to GCC. GCC efficiency on/in a carrier was reduced due to embedded GCC grains in glue (on laminates) or limited water exchange caused by the carrier (in bags). Technical conditions for two of the tested solutions allowed the effective use of GCC.

Keywords Lake restoration · Non-invasive P removal · Calcite · Ground Calcium Carbonate (GCC) · Phosphate · Sorption · Milling

---

D. Pryputniewicz-Flis · J. Idźkowski · Ł. Kozłowicz · G. Brenk
Advanced Pro-Environmental Remediation Solutions Ltd, ul. Poznańska 1, Nielbark, 13306 Kurzętnik, Poland

A. Bańkowska-Sobczak (✉)
Department of Hydraulic Engineering and Applied Geology, Institute of Environmental Engineering, Warsaw University of Life Sciences – SGGW, Nowoursynowska 159 St., 02776 Warsaw, Poland
e-mail: agnieszka_bankowska@sggw.edu.pl

D. Burska
Institute of Oceanography, Faculty of Oceanography and Geography, Universyty of Gdańsk, Av. Marszałka Piłsudskiego 46, 81378 Gdynia, Poland

M. G. Zamparas and G. L. Kyriakopoulos (eds.), *Chemical Lake Restoration*,
https://doi.org/10.1007/978-3-030-76380-0_6

6.1 Introduction

Chemical restoration is a method widely applied in eutrophic lakes aiming at phosphorus (P) inactivation and reduction of P release from sediment by adding P binding agents into the water body. Different compounds and their application methods are used in this technology. Most commonly, coagulants, as alum or iron salts [6, 13, 14, 24, 25, 27] or lanthanum modified bentonite (LMB) [10, 29, 35, 37, 39], are added to the water column to induce phosphate stripping and its subsequent settling to the lake's bottom.

Although these measures often bring expected improvement of the lake's status, they rise numerous concerns due to the fact that chemical compounds are disposed in the aquatic environment. Their application and accumulation in water bodies can lead to increased metal concentration in water [28, 32, 44] and pH changes [36]. Al, Fe and La can get released from sediments due to changing physio-chemical conditions [32, 35], so their concentrations in the lake water in long-term is hard to predict, especially when aging of Fe/Al hydrooxides and LMB will be taken into account; so is the distribution of LMB and alum flocks, as they often accumulate in certain areas of the bottom of the lake, despite even application [10, 32, 46].

Therefore, attempts have been recently made to develop alternate methods for P binding agent application into aquatic ecosystems, which will allow for P sorption onto the agent and its subsequent removal from the water body. As potential solutions, we proposed bags with solid P binding agent inside [5] (similar approach was also further proposed by [49]) and strips coated with P binding agent (the so-called laminates; [2]). Both methods assume application of a solid P sorbing material (placed in bags or coated onto laminates) into the water column for a given time, to allow for P sorption and its subsequent extraction from the water body together with fixed P. Such a measure ensures P inactivation and P removal from the ecosystem as well as eliminates risks related to P biding agent disposal in the aquatic environment. Therefore, we called this approach *non-invasive phosphate removal* [2] and used this term within the current study to describe the general assumptions of this type of applications.

A good candidate for solid P binding agent to be applied using this approach is calcite. Cc is a calcium carbonate ($CaCO_3$) mineral, naturally occurring in the aquatic environment. Phosphate sorption onto Cc involves three types of binding processes: chemisorption on the mineral surface [17, 21, 41], precipitation of calcium-phosphates [41, 47] and co-precipitation proceeding as adsorption and subsequent incorporation of phosphate ions during growth of new Cc crystals [9, 22]. The latter binding process is widely known as self-purification mechanism, as it takes place in hard waters naturally, especially in summer, due to high temperature and CO_2 stripping by intense photosynthesis which reduce $CaCO_3$ solubility [9, 11].

Cc is available in different forms. Natural, powdered Cc-based materials are ground calcium carbonate rocks (GCCs) widely used in many industries. Synthetic calcites are produced by chemical precipitation (precipitated calcium carbonates;

PCCs). Both, GCCs and PCCs, show wide range of physical properties and sorption capacities. Typically, PCCs are much more efficient in phosphate removal as compared to GCCs due to fine granulation, relatively large specific surface area (SSA) and high purity [1, 4, 19]. However, high cost and low availability of PCCs for large scale applications in the environment are factors limiting their common use for lakes restoration purposes [1, 4]. In contrast, GCCs are cost-effective and commonly available materials due to wide distribution of limestone rocks.

Natural powdered calcites (GCCs) have been already used in lakes restoration by their application into the water column [3, 23, 33, 40]. Previously, for lake restoration purposes, calcium was added as lime ($Ca(OH)_2$) to induce Cc precipitation and phosphate sorption [12, 31, 34]. In two lakes, both, $Ca(OH)_2$ and Cc, were applied [33]. However, $Ca(OH)_2$ was actually abandoned, probably because of harmful effects on biota due to high pH caused by its rapid dissolution [30, 31, 50]. Cc was suggested instead as a safe Ca-based P binding agent, as it does not cause rapid and intense pH changes [3, 12, 43].

Due to advantages of Cc as potential P binding agent in eutrophic water bodies and high availability of natural Cc-based materials (GCCs), it seems reasonable to process GCC materials in order to enhance their sorption properties with respect to phosphate. This can be relatively easily achieved by further fragmentation of GCC, to gain finely grained material exhibiting large surface area. For example, Tsai [42] doubled SSA of calcite material through its grinding in planetary ball mill for 1 h. Effects of grinding are largely dependent upon time of grinding and its force [42]. As to our knowledge, effect of calcite grinding on its sorption properties with respect to phosphate has not been studied so far.

Therefore, in this study, we investigated physical and sorption properties of GCC materials obtained by 74 methods of grinding, differing in time, speed and number of milling balls. We further presented our works on potential solutions for non-invasive removal of phosphate from the aquatic ecosystems, which used newly produced GCCs.

6.2 Processing of GCC for Improved Phosphorus Removal

6.2.1 Source Material

Cc-based material used for processing was GCC coming from limestone deposits in Bukova county, Poland, obtained from the company Lhoist Poland. In this material, calcium carbonate constitutes at least 95% (by mass) with trace presence of $MgCO_3$, SiO_2, Fe_2O_3 and Al_2O_3. Mineral composition is dominated by Cc. According to manufacturer data, the particle size of GCC is less than 4 mm. Particles size distribution of GCC determined with sieve analysis (Table 6.1) showed the highest share of particles larger than 0.25 and smaller than 2 mm (64.8%). Grains smaller than 0.25 mm represent 20.4% by mass. Specific granulometric fractions

Table 6.1 Particle size distribution of GCC, surface area and phosphate removal determined for specific fractions Results are means from three replicate experiments

Granulometric fraction (mm)	Mass share (%)	SSA (m^2/g)	Standard phosphate removal rate (%)
>2	14.8	5.98	23.0
2.000 > ϕ > 1.000	25.6	6.44	21.8
1.000 > ϕ > 0.500	21.2	7.32	17.5
0.500 > ϕ > 0.250	18.0	7.37	24.2
0.250 > ϕ > 0.125	11.0	7.76	21.3
0.125 < ϕ > 0.063	5.0	8.64	26.2
<0.063	4.4	7.90	26.7
Weighted	–	7.05	21.9

were subjected to SSA determination using N_2 adsorption method (Autoflow BET+, Quantachrome). SSA ranged from 5.59 m^2/g for the coarse particles (<2 mm) to 8.47 m^2/g for the fine fraction $0.125 < \phi < 0.063$ mm (Table 6.1).

In order to determine phosphate fixation efficiency of GCC, simple sorption test was done for each of the granulometric fractions by shaking (150 rpm, 24h) of 50 mg of each fraction in 1L of an experimental solution mimicking natural lake water. We applied the synthetic lake water in order to closely model in-lake conditions. As sorption processes are strongly affected by solution chemistry, it is important to investigate sorption processes under conditions reflecting target applications [7]. The solution was prepared by dissolving the following salts: $CaCl_2*2H_2O$, $MgSO_4$, KNO_3, $NaHCO_3$ and $KHPO_4$ (292 mg/L; 50 mg/L; 1.6 mg/L; 207 mg/L and 1.4 mg/L, respectively), in double distilled water. pH of the solution was then adjusted to 7.5 by adding 1 M HCl. Phosphate concentration was 1.0 ± 0.1 mg $PO_4{}^{3-}$/L. Sorption test was performed at the temperature of 6 °C in order to model in-lake conditions during spring, when high concentrations of phosphate are usually present in lakes, determining high productivity in the growing season. At the given ionic composition, pH and temperature, the lake water was in equilibrium with respect to Cc (saturation index of −0.1; calculated in PHREEQC software based on solubility product of Cc (log *Ksp)* of 8.48). This solution was used in all the experiments described in this section. Sorption test, as described above, was then multiply repeated for new GCC materials obtained by milling, and was called "standard sorption test", whereas resulting phosphate fixation efficiency was called "standard removal rate". Concentration of phosphate was determined using spectrophotometric molybdenum blue method with ascorbic acid as reducing agent, at wavelength of 880 nm after sample filtration trough 0.45 μm MCE filters. This method of phosphate concentration measurement was applied in all the tests described.

Standard removal rate of individual granulometric fractions of GCC ranged from 17.5% to 26.7%. Thus, taking into account mass share of specific fractions, their SSA and phosphate sorption efficiency, weighted SSA of GCC was calculated as 7.05 m^2/g and weighted standard phosphate removal as 21.9%.

6.2.2 Milling of GCC and Properties of the Obtained Materials

GCC (50 g) was subjected to grinding in a planetary ball mill (Fritsch, Pulverisette 6) using different sets of milling force and time (Table 6.2). Number of milling balls and their diameter ranged from 11 to 123 and from 8 to 15 mm, respectively, which resulted in balls' mass of 150–500 g, accordingly. The range of mill rotation speed and milling time used was 150–400 rpm and 0.5–24 h, respectively. Altogether, 74 different methods of milling were applied. Obtained materials were then investigated in terms of particle size (laser diffraction method after ultrasound treatment for 4 min; Mastersizer—v3.63 Malvern Instruments) and SSA (N_2 adsorption method; Autoflow BET+, Quantachrome).

Milled samples had specific diameter d_{90} ranging from 7.3 to 86.6 μm (Table 6.2). Change in SSA due to processing was substantial, with up to 3.4-fold increase after 18h of milling (Table 6.2). Resulting SSA (up to 24 m^2/g) is relatively high as compared to syntethic calcites (PCCs), for which lower SSA range of 4.2–20 $m^2\ g^{-1}$ was typically reported [1, 15, 18, 19, 45, 47], although much higher values can also be found in the literature (e.g. 67 m^2/g [4]). Increase in SSA was reflected in enhanced phosphate removal (up to 98%), which was up to 4.5-times higher compared to the initial GCC (Table 6.2). Most of the newly produced materials (98% of samples obtained) achieved distinctly higher phosphate removal compared to the initial GCC (Table 6.2).

Substantial increase in SSA (threefold) and in phosphate removal rate (fivefold), compared to the initial material, was obtained already after very short milling time

Table 6.2 Milling parameters and properties of the materials obtained (standard phosphate removal rate determined in the standard sorption test). Results are means from 3 replicate experiments

Milling parameters			Properties of the obtained materials				
Milling time (h)	Ball mass (g)	Rotation speed (rpm)	Specific diamater (μm)			SSA (m^2/g)	Standard phosphate removal rate (%)
			d_{10}	d_{50}	d_{90}		
Initial GCC			n.d.	n.d.	n.d.	7.05	21.9
0.5	150–500	150–550	0.9–2.8	2.1–15.9	7.3–82.5	6.0–20.8	19.5–97.0
1	150–300	200–300	1.1–1.4	2.7–5.1	15.0–86.6	8.1–9.5	27.9–52.1
2	150–250	200–300	1.0–1.4	2.5–3.9	13.2–23.0	6.4–10.5	26.8–63.6
3	150–250	200–300	1.0–1.2	2.7–5.5	16.5–19.2	11.1–11.4	37.4–55.8
4	150–250	200–300	0.9–1.7	2.6–6.5	16.5–37.2	10.7–15.9	46.0–79.5
6	150–250	200–400	0.9–1.7	2.6–12.6	15.7–57.9	8.6–17.6	48.1–93.2
8	150–250	200–300	1.1–1.7	2.7–8.3	12.0–42.9	7.2–19.9	49.3–90.3
12	150–250	150–300	1.0–2.1	2.6–12.1	10.5–57.4	7.4–14.6	32.1–94.3
18	150–250	150–400	1.3–2.8	6.6–14.3	19.3–53.9	5.3–24.0	71.0–98.2
24	150–250	200–300	2.1–2.2	8.1–11.6	32.6–47.9	4.9–7.3	93.0–97.3

(0.5 h). Therefore, in further research we focused on short grinding because it is convenient in terms of time needed to process the material and thus, cost of its production. Particle size diameter, SSA and phosphate removal obtained after short grinding (0.5 h) are presented in Fig. 6.1. After 0.5h of grinding, at rotation speeds from 150 to 400 rpm, increase in mass of milling balls from 150 to 500 g resulted in enhanced SSA and phosphate removal (Fig. 6.1). However, at low rotation speeds of 150 and 200 rpm, both parameters substantially increased only when using the highest mass of balls (500g) (Fig. 6.1). Meaningful increase in SSA and phosphate removal took place also at higher rotation speeds (450–550 rpm), when mass of balls increased from 150 to 250 g. In contrast, further addition of balls (500 g) at these rotation speeds, did not cause a rise in SSA. Moreover, at 500 and 550 rpm, SSA declined. This is most pronounced at the highest rotation speed (550 rpm), where SSA decreased by 37% (from 17.3 to 10.9 m^2/g) when mass of balls was changed from 250 to 500 g (Fig. 6.1). Interestingly, this decline in SSA was accompanied by further increase, albeit slight, in phosphate removal rate (Fig. 6.1).

Our results are in line with those reported by Tsai [42] who investigated effect of milling rotation speed on SSA of Cc-based natural biomaterials (oyster shells) at short grinding time of 0.5 h. He observed similar pattern of SSA change with increasing rotation speed, with clear rise in SSA at rotation speed of up to 400 rpm and further decrease in SSA at higher rotation speeds (500 and 600 rpm). He reported 400 rpm being optimal milling speed at short grinding time (0.5 h) due to the high surface roughness and porosity of the resulting samples, whereas higher rotation speeds led to reduced porosity [42]. Reduction of SSA is often observed after grinding of minerals due to aggregation, which is seen as larger particle size [8, 16, 18, 38]. This effect

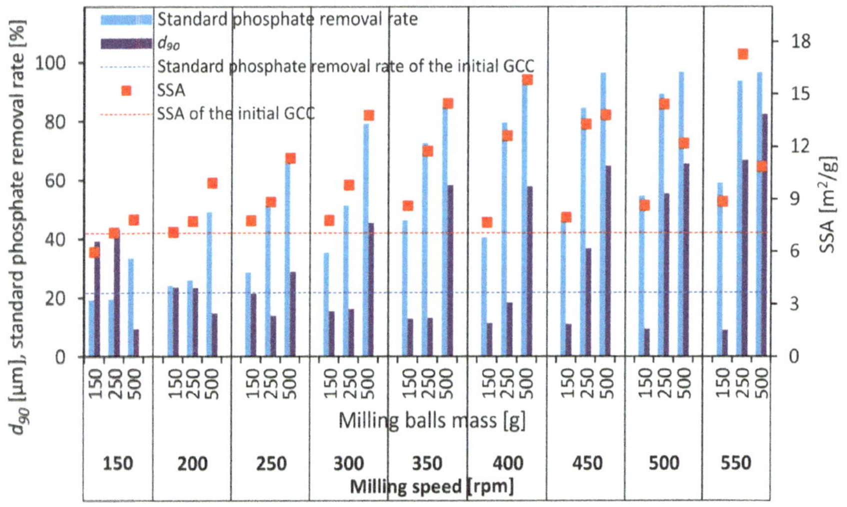

Fig. 6.1 Effect of milling speed and milling balls' mass on SSA, phosphate removal rate and particle diameter (d_{90}) of GCC at milling time of 0.5 h. Results are means from 3 replicate experiments

seems to occur in our samples: at rotation speeds 250–350 rpm, a clear increase in particle size can be observed when using the largest mass of balls (500 g), while at rotation speeds ≥400 rpm, particle size increased also after using medium mass of balls (250 g; Fig. 6.1). However, it is only at high rotation speeds of 500–550 rpm that aggregation leads to a marked decrease in SSA, whereas aggregation does not adversely affect the SSA at rotation speeds of 250–450 rpm (Fig. 6.1). This suggests that aggregates formed at 250–450 rpm are characterized with convenient structure of the surface with numerous pores and slits.

It has to be noted, however, that phosphate removal rates reached here 95%, which is nearly the maximum of what could be achieved. Thus, further investigations at higher phosphate concentration are needed to predict effect of milling on phosphate removal ability of GCC at high rotation speeds and high mass of grinding balls.

As our study showed, by adjusting milling parameters (milling time and speed, number of milling balls), it is possible to produce a Cc-based material characterized by substantially improved phosphate removal ability as compared to the initial GCC, without using chemicals, and at low cost. These new materials were further called "effectively milled".

6.3 Solutions for Non-invasive P Removal from Water Bodies

As described in Sect. 6.1, the main assumption for non-invasive P removal from water bodies is application of solid P binding agent into the water body and its further extraction together with previously sorbed P, to ensure sustain P removal and eliminate non-target effects of the sorbing material on the aquatic environment, as well as risk of desorption. Three different solutions for such application were described in this section. These were: laminates, soft packages (bags) and hard packages (cassettes). The solutions were tested using various Cc-based materials, including effectively milled GCCs. Details on Cc-based materials used and other methodological information are provided in specific subsections.

6.3.1 Laminates

Laminates were prepared as strips consisting of a base material coated with glue and GCC grains (Fig. 6.2). This solution assumes application of laminates into the water column, subsequent phosphate sorption onto Cc layer present on the laminates surface and laminates removal from the water column together with sorbed phosphate.

In the first tests, basalt fabric as laminates' backing and synthetic polyurethane glue were applied (Fig. 6.3). Laminates were coated with two GCC materials: fine (d_{50} = 10 μm) and coarse (d_{50} = 199 μm). Calcium carbonate constituted min.

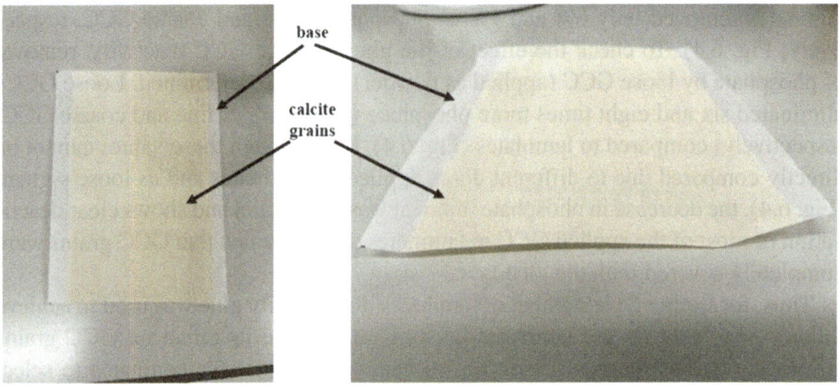

Fig. 6.5 Laminates consisting of light food foil and biodegradable glue coated with GCC

24 g/m^2 of GCC, which is 5–56 less compared to the first laminates made of basalt fabric (Fig. 6.4). Laminates were produced in the Research and Development Centre for Protection and Restoration of Freshwater Ecosystems (Nielbark, Poland).

Microscopic observations showed that particles of slightly coarser GCC material ($d_{50} = 8.6$ μm) formed larger clusters and covered the laminates surface to a higher extent (Figs. 6.6c, d) compared to the finer GCC ($d_{50} = 2.6$ μm) (Figs. 6.6a, b). Moreover, in the laminate containing the finer GCC ($d_{50} = 2.6$ μm), aggregates formed by GCC grains were stuck together by glue and some grains were completely immersed in the glue layer (Figs. 6.6a, b). This suggests that, despite similar GCC mass present in both laminates (24 m^2/g), higher mass of GCC can be exposed (not covered with the glue layer) at the laminates surface when the coarser GCC ($d_{50} = 8.6$ μm) was used.

Macroscopic observations were confirmed in sorption tests conducted for both laminates and both GCC materials (Fig. 6.7). Laminates with slightly coarser GCC material ($d_{50} = 8.6$ μm) showed nearly the same phosphate removal efficiency as compared to loose GCC (Fig. 6.7). This implies that particles of GCC material were well exposed in water and large surface area was available for adsorbing phosphate. However, laminates with the finer GCC material ($d_{50} = 2.6$ μm) removed 30% less phosphate compared to loose GCC (Fig. 6.7) showing high deactivation of GCC particles, most likely due to aggregation and coverage by glue, as previously observed (Fig. 6.6).

The modified laminates (foil and low-density biodegradable glue layer) showed much smaller deactivation of GCC material compared to the initial laminates (basalt fabric and polyurethane glue) (Table 6.3). The lowest and negligible deactivation (1%) was shown by the modified laminate containing GCC with median particle size (d_{50}) of 8.6 μm.

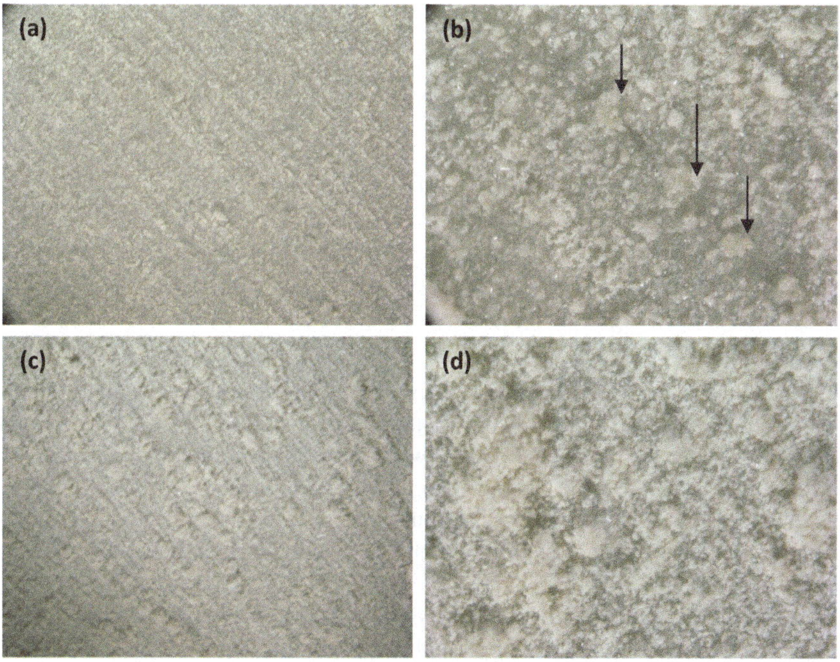

Fig. 6.6 GCC layer onto laminates: GCC with granulation of $d_{50} = 2.6\ \mu m$ ($d_{10} = 1.0\ \mu m$, $d_{90} = 12.6\ \mu m$) at magnification 10 x (**a**) and 50 x (**b**) and GCC with granulation of $d_{50} = 8.6\ \mu m$ ($d_{10} = 1.5\ \mu m$, $d_{90} = 54.7\ \mu m$) at magnification 10 x (**c**) and 50x (**d**). Stereoscopic microscope NIKON SMZ 1500. Arrows indicate particle aggregates immersed in the glue layer (photos by Dorota Pryputniewicz-Flis)

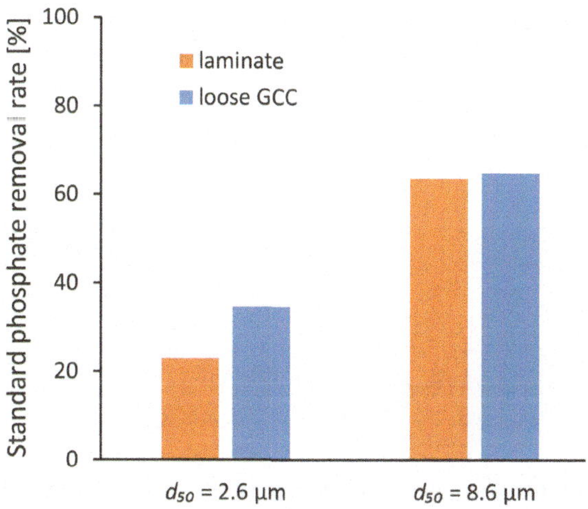

Fig. 6.7 Phosphate removal by laminates with GCCs and by loose GCCs (granulation of $d_{50} = 2.6\ \mu m$ and $d_{50} = 8.6\ \mu m$). Mass of GCC in laminates and applied as loose material was the same (0.2 g/L). Laminates and loose GCC were shaken for 24 h (rocker shaker BIOMETRA WT16, 40 cycles/min.) in artificial lake water (see Sect. 6.2.1) at room temperature and initial phosphate concentration of 1 mg PO_4^{3-}/L. Results are means from 3 replicate experiments

Table 6.3 Deactivation of GCC materials in two types of laminates. Deactivations refers to difference between phosphate removal by loose GCC (applied as powder) and by laminates containing the same GCC. Results are means from 3 replicate experiments

Laminates type	Cc material (μm)	Deactivation (%)
Basalt fabric/polyurethane glue	GCC, $d_{50} = 10$	85
	GCC, $d_{50} = 199$	86
Food package foil/biodegradable glue	GCC, $d_{50} = 1.6$	30
	GCC, $d_{50} = 8.6$	1

6.3.2 *Soft Packages (Bags)*

Further solution considered for non-invasive removal of P from the aquatic environment was a soft package (bag) filled with a Cc-based material (Fig. 6.8). The general assumption is that the package is made of water permeable fabric which allows water to come into contact with the sorbing material and prevents its loss to the surrounding medium [5]. Burska et al. [5] tested this solution for two Cc-based materials. A similar approach (called "tea-bags") was further investigated by Zamparas et al. [49] for two modified bentonites.

First tests for Cc application in bags were performed using two Cc-based materials: GCC ($d_{50} = 10$ μm; the same as applied in modified laminates, see Sect. 6.3.1) and PCC (synthetic, pure Cc; $d_{50} = 0.08$ μm; specific diameter and surface area as reported by [1]. Bags were placed in natural lake water (previously filtered through filter paper) for 48 h. The dosage of both Cc materials was 1 g/L. Concentration of phosphate substantially dropped within 24 h. During continued incubation, it remained unchanged (GCC; $d_{50} = 10$ μm) or slightly decreased (PCC; $d_{50} =$ 0.08 μm) (Fig. 6.9). Bags with PCC were much more efficient in phosphate sorption compared to GCC (after 24 h, corresponding phosphate removal accounted for 35% and 16%; Fig. 6.9). Mass of the removed phosphate was 0.04 mg PO_4^{3-}/g (GCC) and

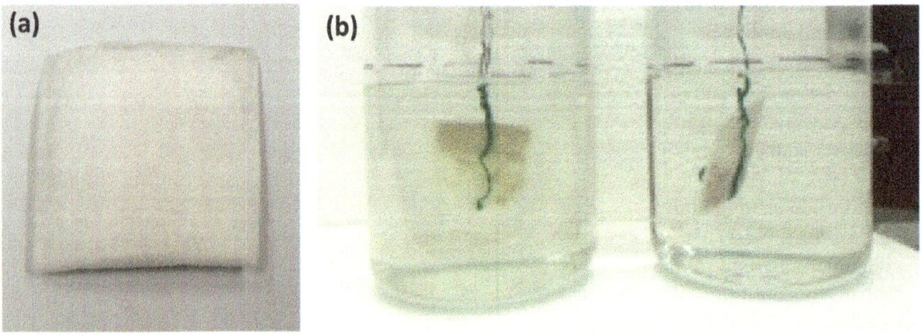

Fig. 6.8 Bags filled with Cc-based material: **a** bag with PCC; **b** bags placed in water (Cc material applied is GCC (left) and PCC (right); 1 g/L)

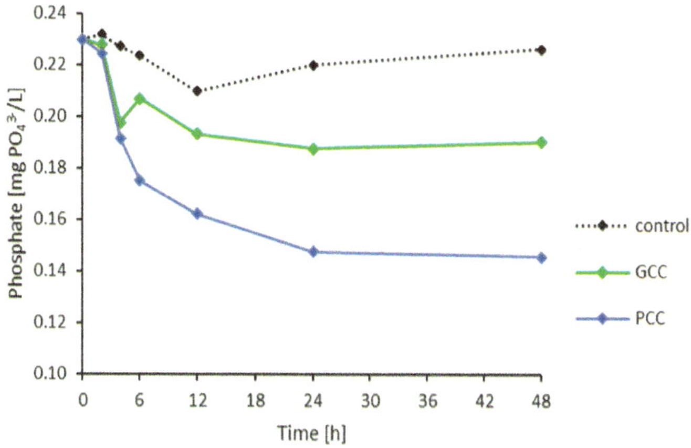

Fig. 6.9 The concentration of phosphate after 2, 4, 6, 12, 24 and 48 h in the presence of bags filled with Cc-based materials (GCC (d_{50} = 10 μm) and PCC (d_{50} = 0,08 μm)) and in control (no bag). Natural lake water (Lake Stężyckie, north Poland) filtered through filter paper; initial phosphate concentration = 0.23 mg PO_4^{3-}/L. Experiment was performed at the temperature of 20 °C. Water was mixed with the glass rod only prior to sampling (no continuous mixing). Results are means from 3 replicate experiments

0.08 mg PO_4^{3-}/g (PCC). Difference in phosphate fixation between the two Cc materials reflect their physical properties—grain size and surface area—which determine surface sorption capacity [1, 4].

Interestingly, some minor changes in phosphate concentration were also recorded in control (no bags, no sorbing material) (Fig. 6.9). This depletion of phosphate was taken into account when calculating phosphate removal by bags. Clear decrease in phosphate concentration in control could result from several processes: sorption of phosphate onto organic and inorganic matter (which were probably present in water despite filtration) or assimilation of phosphate by microorganisms. Thus, total phosphorus (P_{tot}) was also determined (Fig. 6.10). Depletion of P_{tot} indicates that not only phosphate, but also other phosphorus species must have been fixed by bags filled with Cc-based materials (Fig. 6.10). It is likely that, beside phosphates, polyphosphates were removed, as Cc is able to sorb these compounds, too [26].

When considering total phosphorus pool in the removal balance, it becomes clear that efficiency of bags should be higher, especially of these with PCC (by about 30%). However, P removal efficiency recorded in the experiment is rather low. Burska et al. [5] assessed that bags reduced P sorption by about 84% (GCC) and 76% (PCC) as compared to loose calcites. This effect could result from limited penetration of water into the bags due to the clogging ot the material pores by GCC particles. Zamparas et al. [49] observed that bags with lanthanum modified benthonite (LMB) removed phosphate much slower as compared to loose LMB. To achieve similar level of phosphate removal, longer exposure of bags in the experimental solution was required. This indicates that bags delayed contact of water with the sorbent. In our experiment with bags, phosphate sorption was probably hampered after 48 h.

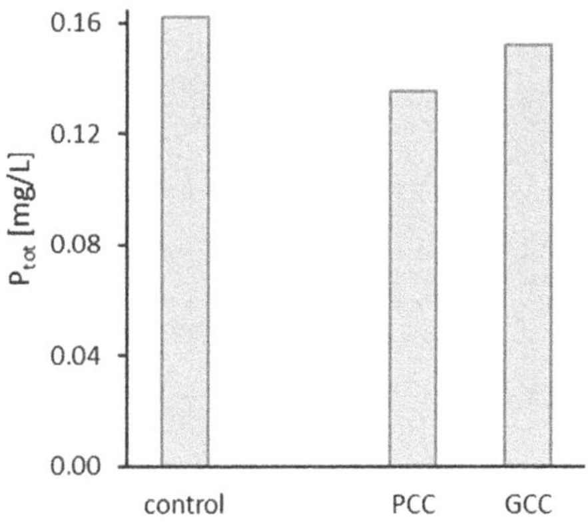

Fig. 6.10 Concentration of total phosphorus (P_{tot}) in the lake water after 48 h in the presence of bags filled with Cc-based materials (GCC (d_{50} = 10 μm) and PCC (d_{50} = 0,08 μm)) and in control (no bag). The dose of both Cc materials was 1 g/L. Experimental conditions as described in Fig. 6.9. Results are means from 3 replicate experiments. Total phosphorus was determined after sample digestion at elevated temperature and pressure in acidic environment with potassium persulfate as an oxidizing agent

However, at more intense (continuous) mixing, higher phosphate removal success would be probably achieved.

Another issue important for bags application is the possible loss of sorbent due to the fact that some grains may fall out through pores in the packaging. During incubation experiment in the lake water, no visible loss of GCC/PCC was recorded, most probably due to rather static conditions prevailing in the beakers (no continuous mixing). However, when removing the bags from the beakers, we observed that some Cc material got out through the pores in the packaging fabric. Thus, further investigations on the bags were performed under continuous but slight water mixing (rocker shaker; BIOMETRA WT16; 40 cycles/min.). Moreover, attempts were made to optimize selection of packaging material to ensure possibly high and fast water penetration into the bags without substantial Cc loss. For this purpose, filter papers laminated with nonwovens were prepared by Institute of Biopolymers and Chemical Fibres (Łukasiewicz Research Network, Łódź, Poland). Four types of filter papers were tested in two variants of adhesive coating (Table 6.4). The bags were filled with a Cc-based material obtained by effective milling of GCC (milling time of 0.5 h at 350 rpm; SSA = 14.1 m^2/g, d_{50} = 8.6 μm). This GCC was further used in all experiments described in the current study. The dose of GCC was 1 g/L (1 g of GCC in each bag (or as loose sorbent) incubated in 1L of the experimental solution).

In the first variant of the glue coating, there was only a minor loss of GCC material from the bags. However, at the same time, very low phosphate removal was recorded for these type of packages, indicating that the glue layer substantially hampered water penetration into the bags (Table 6.4). After 24 h of incubation with bags, depletion of phosphate did not exceed 14% (phosphate mass removed per unit mass of GCC

Table 6.4 Phosphate removal in the presence of bags filled with GCC (d_{50} = 8.6 μm). Bags were made of filter papers laminated with nonwoven in two variants of coating with glue. Bags with GCC were shaken for 24 h (rocker shaker BIOMETRA WT16, 40 cycles/min.) in artificial lake water (see Sect. 6.2.1) at room temperature and initial phosphate concentration of 1 mg PO_4^{3-}/L. Results are means from 3 replicate experiments

Bags		Phosphate removal (%)	Loss of GCC from bags (%)
Glue coverage variant	Filter paper type		
I	1	11	≤2.13
	2	14	≤1.68
	3	12	≤0.11
	4	10	≤1.97
II	1	51	≤25
	2	48	≤23
	3	35	≤15
Loose GCC		87	–

was 0.05–0.12 mg PO_4^{3-}/g), whereas loose GCC eliminated 87% of phosphate (0.96 mgPO_4^{3-}/g).

In the second variant of the glue coating (less glue), the packages released 15–25% of GCC mass initially placed in bags, and a higher phosphate fixation was achieved (35–51%; Table 6.4). Thus, it can be concluded that the higher sorption of phosphate was mainly due to presence of loose sorbent in the water and, to a lesser extent, due to better water penetration into bags.

Further attempts to modify the construction of bags made of filter papers laminated with nonwoven did not bring expected improvements. Therefore, other materials were tested as candidates for bags making. Commercially available agricultural nonwovens, filter materials and technical fabrics were tested in terms of GCC loss and water penetration into bags and its effect on phosphate sorption. Similar problems, as previously described, occurred. Either the contact of GCC with water and phosphate sorption were limited (to low permeability of the packaging) or there was substantial loss of GCC from the bag (to high permeability of the packaging). In the latter case, recorded phosphate removal was as high as 90%, but phosphate sorption took place mainly on the loose portion of GCC released from the bag (and not on GCC remaining in the bag). Moreover, we observed that wet GCC tends to clump and fall onto bottom of the package. We used these properties of wet GCC to construct another solution for Cc application in eutrophic water bodies without its disposal in aquatic environment (hard packages, Sect. 6.3.3).

6.3.3 *Hard Packages (Cassettes)*

This solution is based on tray-like vessels in which Cc-based material is placed. From the top, the trays are covered with a dense net to prevent Cc material grains

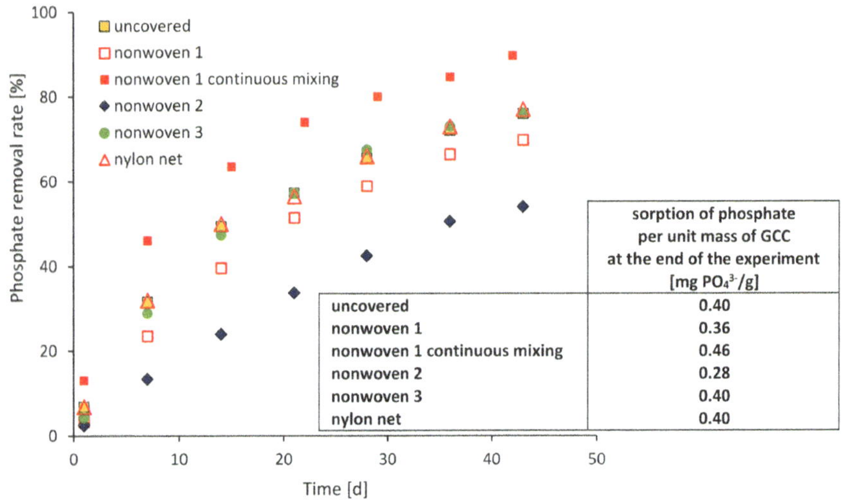

Fig. 6.13 Phosphate removal during cassettes incubation in artificial lake water (see Sect. 6.2.1) at room temperature and initial phosphate concentration of 1 mg PO_4^{3-}/L. Cassettes contained effectively milled GCC (d_{50} = 8.6 μm, 25 g in each, which corresponds to GCC dosage of ~ 2 g/L) placed as merged layer, and were covered with nonwovens [1–3], nylon net or left uncovered. Water was mixed once a day with a glass rod in all the series except for nonwoven 1 for which additional experiment with continuous mixing was also performed (water flow velocity was about 10 cm/s). Results are means from 3 replicate experiments

to the same cassette incubated without continuous mixing, although the difference dropped to 1.7– times on the last day of the experiment. Moreover, the covered cassette under continuous mixing caused 1.2–1.9 times higher phosphate depletion as compared to the uncovered cassette incubated without continuous mixing.

These results showed that water flow above the cassettes is an important factor affecting phosphate removal efficiency by GCC. Water flow provides continuous phosphate transport into the cassette and enables its contact with surface of the sorbent as well as prevents formation of chemocline which could hamper diffusion of phosphate. Such dynamic conditions are expected to occur in the lake during cassettes application, exposure and removal.

In the experiments previously described, cassettes were exposed to a defined volume of water, so concentration of phosphate gradually decreased over time. Under in-lake conditions, persistent high concentrations of phosphate can occur, especially in autumn and winter (lack of intense primary production) and at near bottom depths (P release from sediment). Hence, the water flow and diffusion processes will maintain phosphate supply into the cassette and the decline in phosphate concentration will not be as steep as observed in these experiments (limited mass of phosphate present). Therefore, further investigations of cassettes' performance employed dynamic flow conditions with continuous water exchange. Moreover, in the next step, we also checked the effect of GCC dose applied. In previous experiments, GCC dose applied per 1L of the experimental solution was 1 g (Fig. 6.12) and 2 g (Fig. 6.13), which corresponded to GCC layer thickness of 0.04 g/cm^2 and 0.18 g/cm^2, respectively (area of the cassettes' bottom was 28.3 and 132.7 cm^2 and water volume was 1L and 12 L, respectively). Based on the dynamics of phosphate depletion in these experiments, it can be predicted that cassettes with thick GCC layer (0.18 g/cm^2) can achieve comparable unit sorption (0.9 mg PO_4^{3-}/g) as compared to cassettes with thin GCC layer (0.04 g/cm^2), after ten times longer time of exposure.

In dynamic flow experiments, water exchange was employed using a peristaltic pump (0.5 L/day; WATSON MARLOW) and water movement was obtained by continuous mixing (rocker shaker, 40 cycles/min.) (Fig. 6.14). Cassettes (petri plates, diameter of 6 cm), uncovered, were incubated with 1L of artificial lake water (see Sect. 6.2.1), provided and exchanged by the peristaltic pump for 210 days. GCC (d_{50} = 8.6 μm) was applied in four doses: 0.5, 1.0 and 5 and 10 g/L (corresponding mass of GCC in the cassettes were 0.02, 0.04, 0.18 and 0.35 g/cm^2). The effluent was sampled for phosphate determination.

After application of small GCC doses (0.02 and 0.04 g/cm^2), phosphate removal actually ceased after about two months (70 days) (Fig. 6.15). Thus, no further observations were conducted for these series of the experiment. The cumulative unit sorption was high and accounted for 9.56–12.42 mg PO_4^{3-}/g (70 days, Table 6.5). After the same period of cassettes incubation with higher GCC doses (0.18 and 0.35 g/cm^2), unit sorption was much lower and accounted for 2.58 and 1.44 mg PO_4^{3-}/g, respectively (Table 6.5). Longer exposure of these cassettes in the experimental solution

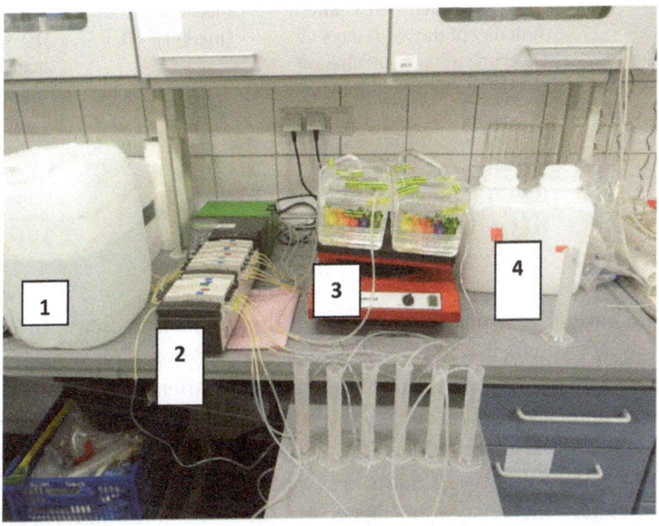

Fig. 6.14 Experimental setup with water exchange and continuous mixing: (1) container for fresh artificial lake water (see Sect. 6.2.1) with initial phosphate concentration of 1 mg PO_4^{3-}/L; (2) peristaltic pump, (3) rocker shaker (BIOMETRA WT16; 40 cycles/min) with incubation chambers containing cassettes with GCC (d_{50} = 8.6 μm); (4) containers for effluent. The image presents control flow measurement: water was sampled to cylinders in a given time unit to control water flow rate. Experiments were performed at room temperature

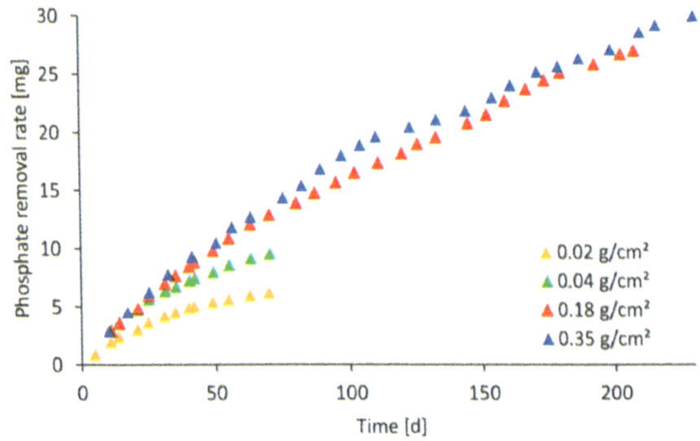

Fig. 6.15 Cumulative phosphate removal from artificial lake water (see Sect. 6.2.1) under water exchange and continuous mixing, at room temperature. Initial phosphate concentration was of mg PO_4^{3-}/L). Four GCC (d_{50} = 9.6 μm) doses (0.02–0.35 g/cm^2) were tested

Table 6.5 Phosphate concentration in the effluent and phosphate sorption after 70 and 210 days of the experiment with cassettes under water exchange and continuous mixing (Fig. 6.14). The experiment was performed in artificial lake water (see Sect. 6.2.1.) at initial phosphate concentration of 1 mg PO_4^{3-}/L and room temperature. Cassettes contained GCC (d_{50} = 8.6 μm) in four doses. Results are means from 3 replicate experiments. For low GCC dosages (0.02 and 0.04 g/cm^2) phosphate removal ceased after 70 days so no further observations were made for these variants of GCC mass applied

GCC dose (g/L)	GCC mass in the cassette (g/cm^2)	Phosphate concentrations (range) in effluent between 1st and 70th day of the experiment (mg PO_4^{3-}/L)	Cumulative sorption of phosphate per unit mass of GCC after 70 days of incubation (mg PO_4^{3-}/g)	Maximum phosphate concentration after 210 days of incubation (mg PO_4^{3-}/L)	Cumulative sorption of phosphate per unit mass of GCC after 210 days of incubation (mg PO_4^{3-}/g)
0.5	0.02	0.64–0.95	12.42	–	–
1	0.04	0.59–0.91	9.56	–	–
5	0.18	0.59–0.78	2.58	0.94	5.40
10	0.35	0.45–0.74	1.44	0.82	2.91

resulted in a further phosphate removal and almost linear increase in cumulative removal (Fig. 6.15) which achieved 2.91 and 5.40 mg PO_4^{3-}/g, respectively, after 210 days (Table 6.5). At the end of the experiment (after 210 days), the cassettes still removed phosphates (Fig. 6.15), however, reduction of phosphate concentration in this stage of the experiment was low (6%) which is shown by high phosphate concentration in the effluent, especially in case of GCC dose of 0.18 g/cm^2 (Table 6.5).

Taking into account the sorption capacity of the sorbent, low mass of GCC added to the cassettes (0.02 g/cm^2) seems to be the optimum solution (the highest P sorption per unit mass of GCC). Multiple but low GCC dosages should be applied as sets containing number of single cassettes. Cassettes with large doses of GCC (e.g. 0.35 g/cm^2) can be considered as a kind of "loading dose", when high phosphate concentrations are to be fast reduced. Further exposure of the cassettes will result in continuous phosphate sorption on the remaining GCC surface. For complete utilization of the sorbents' capacity, the exposure should last several months. Adjustment of GCC dose (within the volume of a single cassette) is an important advantage of this solution.

6.4 Summary

The presented solutions for non-invasive P removal from aquatic ecosystems using processed GCC differed in the construction, mass of GCC possible to be applied and efficiency of P removal.

Soft packages (bags) are in theory easy to make and apply n. However, proper selection of the P sorbing agent (in terms of grain size) and packaging material (in terms of water permeability) to achieve high P removal and high retention of the sorbent in the bag, remains a challenge.

Laminates require more advanced works to be produced, and amount of P sorbing agent which can be applied per unit area, is relatively small. This solution can be regarded as a short-term measure (to be applied for short periods, such as days), as relatively small mass of sorbent (24 g/m^2) can be fixed on their surface, so their sorption capacity is limited. Therefore, repeated applications of the laminates for short periods should be considered rather than longer exposure in the environment. However, this needs to be verified in long-term experiments. The problem of the relatively low mass of GCC present in the laminates, can be naturally overcome by adjusting surface size of the laminates' and their quantity. However, based on the experimental results, we calculated that the laminates area required to completely remove phosphate from 1 m^3 of the lake water (assuming an initial concentration of 1 mg PO_4^{3-}/L and the same Cc-based material used) is quite large and accounts for at least 8.1 m^2 (Table 6.6). On the other hand, the area (and thus number of laminate strips needed) can be reduced twice by double-sided coating with the P binding agent. This direction seems to be reasonable. Assuming homogeneous

Table 6.6 Estimated area of laminates and mass of Cc-based material needed for complete phosphate removal from 1 m^3 of the lake water at initial phosphate concentration of 1 mg PO_4^{3-}/L (calculations based on previous experiments for laminates made of food package foil as base and biodegradable glue)

Cc-based material applied	Laminate's area (m^2)	Sorbent mass needed (g)
GCC (d_{50} = 2.6 μm)	22.5	540
GCC (d_{50} = 8.6 μm)	8.14	195

coating of the laminate on both sides, the sorbent mass applied on one laminate strip could be doubled.

For longer exposure in the water body, cassettes seem to be a more suitable solution. The mass of GCC which can be placed in the cassettes is relatively large (in our tests—200–3500 g/m^2, Table 6.5, which is 8–145 more compared to laminates (24 g/m^2 at one-side coverage)) and thus it is sufficient to ensure substantial P removal for few to dozen weeks. For example, the cassettes with the lowest GCC dose (0.02 g/cm^2) would be able to completely remove phosphate from 1 m^3 of the lake water (assuming an initial concentration of 1 mg PO_4^{3-}/L) within 14 days at their total area of 0.952 m^2 (total mass of GCC applied is 191 g). If this effect would have to be achieved within 70 days, two times smaller area of cassettes and mass of GCC would be needed (0.402 m^2 and 81 g, respectively; Table 6.7). The required surface area of the cassettes could be reduced by adding a larger dose of GCC. For example, by assumed application for 70 days to achieve complete phosphate elimination, the surface area of the cassettes would drop to 0.192 m^2 at GCC dosage of 0.35 g/cm^2. However, total mass of GCC needed would rise to 692 g and its sorption capacity would be fully utilized after more than six months.

High dosage of GCC in single cassettes can be also used as a kind of "loading dose" to achieve fast (within few days) and substantial phosphate elimination. However, within such a short time, sorption capacity of GCC mass applied would not be fully saturated, so a longer exposure should be considered to allow for further sorption, even if the phosphate elimination after a longer period would be minor.

Thus, the cassettes as a solution for non-invasive P removal, are very convenient as their application parameters (single and total mass of GCC, total surface area of cassettes) are easy to adjust to specific in-lake conditions. In practice, sets of cassettes, placed vertically one above the other, should be considered for field application to allow phosphate sorption at different depths.

Non-invasive removal of P, as described in this study, can (at least in theory) enable P recycling by reuse of the once used sorbent (enriched with phosphates which was previously removed from the water column). Zamparas et al. [49] suggested reuse of modified bentonites (after their application in bags for P removal from water bodies)

Table 6.7 Estimated minimum surface area of the cassettes and mass of Cc-based material needed for complete phosphate removal from 1 m^3 of the lake water at an initial phosphate concentration of 1 mg PO_4^{3-}/L (calculations based on the previous experiments for GCC (d_{50} = 8.6 μm))

Mass of GCC in the cassette (g/cm^2)	Exposure time (d)			
	14		70	
	Area of cassettes (m^2)	GCC mass required (g)	Area of cassettes (m^2)	GCC mass required (g)
0.02	0.952	191	0.402	81
0.04	0.676	270	0.262	105
0.18	0.751	1351	0.215	388
0.35	0.690	2416	0.198	694

as fertilizers. Among solutions presented in the current study, cassettes and bags seem to be the most suitable for such a measure as sorbent extraction from the cassettes/bags should be relatively easy. In the case of laminates, separation of P binding agent from the base can be more challenging. However, application of a sorbent together with biodegradable glue (which is also present in laminates and would be probably hard to separate from the sorbent itself) seems to be a reasonable solution. If Cc is used as sorbent, recycled mineral enriched with phosphate could be applied for fertilization and soil pH correction. Nevertheless, it has to be remembered that Cc is also able to sorb a number of potentially harmful elements and substances, such as arsenates [41], selenates and selenites [20], cadmium and lead [48]. Therefore, its further reuse in the environment should be carefully considered if Cc was previously applied for restoration of potentially polluted water bodies, such as urban lakes.

Funding Information This research was financially supported by The National Centre of Research and Development No GEKON2/03/267948/21/2016 and European Union's and Republic of Poland within the Smart Growth Operational Programme 2014–2020, Priority axis I: Support for R&D activity of enterprises, R&D projects of enterprises, Industrial Research and Development Works (grant number POIR.01.01.01-00-0981/17-00).

References

1. Bańkowska-Sobczak A, Blazejczyk A, Eiche E, Fischer U, Popek Z (2020.) Phosphorus inactivation in lake sediments using calcite materials and controlled resuspension—mechanism and efficiency. Minerals 10(3):223. https://doi.org/10.3390/min10030223
2. Bańkowska-Sobczak A, Brenk G, Burska D, Idźkowski J, Kozłowicz Ł, Kupiec JM, Powałowski S, Pryputniewicz-Flis D, Sklepik M (2019) Non-invasive phosphate removal from the lake water—principles, efficiency and non-target effects. In: Book of abstracts—11th symposium for European freshwater sciences, June 30–July 5, 2019, Zagreb, Croatia, p. 173. http://www.sefs11.biol.pmf.hr/book-of-abstracts/l. Accessed Sept 20
3. Berg U (2001) Die Kalzitapplikation als interne Restaurierungsmaβnahme für eutrophierte Seen – ihre Optimierung und Bewertung. Technische Universität Kalsruhe (niepublikowana), Rozprawa doktorska
4. Berg U, Neumann T, Donnert D, Nüesch R, Stüben D (2004) Sediment capping in eutrophic lakes—efficiency of undisturbed calcite barriers to immobilize phosphorus. Appl Geochem 19:1759–1771
5. Burska D, Pryputniewicz-Flis D, Bańkowska-Sobczak A, Brenk G, Woszczyk T (2019) The Efficiency of P-Removal from Natural Waters with Sorbents Placed in Water Permeable Nonwovens. IOP Conf Series Earth Environ Sci 362(1):1–10. https://doi.org/10.1088/1755-1315/362/1/012099
6. Cooke GD, Welch EB, Peterson S, Nichols SA (2005) Restoration and management of lakes and reservoirs. CRC Press, Boca Raton. https://doi.org/10.1002/rrr.3450090207
7. Cucarella V, Renman G (2009) Phosphorus sorption capacity of filter materials used for on-site wastewater treatment determined in batch experiments—a comparative study. J Environ Qual 38:381–392
8. Čavajda V, Uhlík P, Derkowski A, Čaplovičová M, Madejová J, Mikula M, Ifka T (2015) Influence of grinding and sonication on the crystal structure of talc. Clays Clay Miner 63(4):311–327

9. Danen-Louwerse HJ, Lijklema L, Coenraats M (1995) Coprecipitation of phosphate with calcium carbonate in Lake Veluwe. Water Res 29(7):1781–1785. https://doi.org/10.1016/0043-1354(94)00301-M
10. Dithmer L, Nielsen UG, Lürling M, Spears BM, Yasseri S, Lundberg D, Moore A, Jensen ND, Reitzel K (2016) Responses in sediment phosphorus and lanthanum concentrations and composition across 10 lakes following applications of lanthanum modified bentonite. Water Res 97:101–110. https://doi.org/10.1016/j.watres.2016.02.011
11. Dittrich M, Koschel R (2002) Interactions between calcite precipitation (natural and artificial) and phosphorus cycle in the hardwater lake. Hydrobiologia 469:49–57. https://doi.org/10.1023/A:1015571410442
12. Dittrich M, Gabriel O, Rutzen C, Koschel R (2011) Lake restoration by hypolimnetic $Ca(OH)_2$ treatment: impact on phosphorus sedimentation and release from sediment. Sci Total Environ 409:1504–1515. https://doi.org/10.1016/j.scitotenv.2011.01.006
13. Dondajewska R, Gołdyn R, Kowalczewska-Madura K, Kozak A, Romanowicz-Brzozowska W, Rosińska J, Budzyńska A, Podsiadłowski S (2020) Hypertrophic lakes and the results of their restoration in Western Poland. In: Korzeniewska E, Harnisz M (eds) Polish river basins and lakes—Part II. The handbook of environmental chemistry, vol 87. Springer, Cham, pp 373–399. https://doi.org/10.1007/978-3-030-12139-6_17
14. Dunalska J, Wiśniewski G (2016) Can we stop the degradation of lakes? Innovative approaches in lake restoration. Ecol Eng 95:714–722. https://doi.org/10.1016/j.ecoleng.2016.07.017
15. Eiche E, Berg U, Song Y, Neumann T (2008) Fixation and phase transformation of phosphate at calcite surfaces—implications for eutrophic lake restoration. In: Proceedings of ninth international congress for applied mineralogy, Brisbane, pp 1–10
16. Emami AH, Bafghi MS, Vahdati Khaki J, Zakeri A (2009) The effect of grinding time on the specific surface area during intensive grinding of mineral powders. Iranian J Mater Sci Eng 6(2):30–36
17. Freeman JS, Rowell DL (1981) The adsorption and precipitation of phosphate onto calcite. J Soil Sci 32:75–84
18. Gammage RB, Glasson DR (1976) The effect of grinding on the polymorphs of calcium carbonate. J Colloid Interface Sci 55(2):396–401
19. Hart B, Roberts S, James R, Taylor J, Donnert D, Furrer R (2003) Use of active barriers to reduce eutrophication problems in urban lakes. Water Sci Technol 47(7–8):157–163. https://doi.org/10.2166/wst.2003.0684
20. Heberling F, Bosbach D, Eckhardt J-D, Fischer U, Glowacky J, Haist M, Kramar U, Loos S, Müller HS, Neumann T, Pust C, Schäfer T, Stelling J, Ukrainczyk M, Vinograd V, Vučak M, Winkler B (2014) Reactivity of the calcite-water-interface, from molecular scale processes to geochemical engineering. Appl Geochem 45(1):158–119
21. Hinedi ZR, Goldberg S, Chang AC, Yesinowski JP (1992) A 31P and 1H MAS NMR study of phosphate sorption onto calcium carbonate. J Colloid Interface Sci 152(1):141–160
22. House WA, Donaldson L (1986) Adsorption and coprecipitation of phosphate on calcite. J Colloid Interface Sci 112(2):309–324
23. Hupfer M (2004) Bedeutung Sedimentstratigraphischer Untersuchungen für die Seentherapie. Studia Quaternaria 21:171–178. http://www.studia.quaternaria.pan.pl/pdfs/sq21/s_171_178.pdf. Accessed Aug 25 20
24. Hupfer M, Hilt S (2008) Lake restoration. In: Jørgensen SE, Fath BD (eds) Encyclopedia of ecology. Elsevier, Amsterdam, pp 2080–2093
25. Huser BJ, Egemose S, Harper H, Hupfer M, Jensen H, Pilgrim KM, Reitzel K, Rydin E, Futter M (2016) Longevity and effectiveness of aluminum addition to reduce sediment phosphorus release and restore lake water quality. Water Res 97:122–132. https://doi.org/10.1016/j.watres.2015.06.051
26. Karageorgiou K, Paschalis M, Anastassakis GN (2007) Removal of phosphate species from solution by adsorption onto calcite used as natural adsorbent. J Hazard Mater 139(3):447–452. https://doi.org/10.1016/j.jhazmat.2006.02.038

27. Kuster AC, Kuster AT, Huser BJ (2020) A comparison of aluminum dosing methods for reducing sediment phosphorus release in lakes. J Environ Manage 261: https://doi.org/10.1016/j.jenvman.2020.110195
28. Łopata M, Augustyniak R, Grochowska J, Parszuto K, Tandyrak R, Wiśniewski G (2020) Behavior of aluminum compounds in soft-water lakes subjected to experimental reclamation with polyaluminum chloride. Water Air Soil Pollut 231:358. https://doi.org/10.1007/s11270-020-04708-6
29. Meis S, Spears BM, Maberly SC, O'Malley MB, Perkins RG (2012) Sediment Amendment with Phoslock® in Clatto Reservoir (Dundee, UK): investigating changes in sediment elemental composition and phosphorus fractionation. J Environ Manage 93:185–193. https://doi.org/10.1016/j.jenvman.2011.09.015
30. Miskimmin B, Donahue W, Watson D (1995) Invertebrate community response to experimental lime ($Ca(OH)_2$) treatment of an eutrophic pond. Aquat Sci 57(1):20–30. https://doi.org/10.1007/BF00878024
31. Murphy TP, Hall KJ, Yesaki I (1983) Coprecipitation of phosphate with calcite in a naturally eutrophic lake. Limnol Oceanogr 28(1):58–69. https://doi.org/10.4319/lo.1983.28.1.0058
32. van Oosterhout F, Waajen G, Yasseri S, Manzi Marinho M, Pessoa Noyma N, Mucci M, Douglas G, Lürling M (2020) Lanthanum in Water, Sediment, Macrophytes and chironomid larvae following application of Lanthanum modified bentonite to lake Rauwbraken (The Netherlands). Sci Total Environ 706: https://doi.org/10.1016/j.scitotenv.2019.135188
33. Prepas EE, Pinel-Alloul B, Chambers PA, Murphy TP, Reedyk S, Sandland G, Serediak M (2001) Lime treatment and its effects on the chemistry and biota of hardwater eutrophic lakes. Freshw Biol 46(8):1049–1060. https://doi.org/10.1046/j.1365-2427.2001.00788.x
34. Reedyk S, Prepas EE, Chambers PA (2001) Effects of single $Ca(OH)_2$ doses on phosphorus concentration and macrophyte biomass of two boreal eutrophic lakes over 2 years. Freshw Biol 46(8):1075–1087
35. Reitzel K, Andersen FT, Egemose S, Jensen HS (2013) Phosphate adsorption by lanthanum modified bentonite clay in fresh and brackish water. Water Res 47:2787–2796. https://doi.org/10.1016/j.watres.2013.02.051
36. Reitzel K, Jensen HS, Egemose S (2013) pH dependent dissolution of sediment aluminum in six Danish lakes treated with aluminium. Water Res 47(3):1409–1420
37. Robb M, Greenop B, Goss Z, Douglas G, Adeney J (2003) Application of Phoslock™, an innovative phosphorus binding clay, to two Western Australian waterways: preliminary findings. Hydrobiologia 494(1–3):237–243
38. Sánchez-Soto PJ, Justo A, Pérez-Rodríguez JL (1994) Structural alteration of pyrophyllite by dry grinding as studied by IR spectroscopy. J Mater Sci Lett 13:915–918
39. Spears BM, Lürling M, Yasseri S, Castro Castellon AT, Gibbs M, Meis S, McDonald C, McIntosh J, Sleep D, van Oosterhout F (2013) Lake responses following lanthanum-modified bentonite clay (Phoslock®) application: an analysis of water column lanthanum data from 16 case study lakes. Water Res 47(15):5930–5942
40. Stüben D, Walpersdorf E, Voss K, Rönicke H, Schimmele M, Baborowski M, Luther G, Elsner W (1998) Application of lake marl at Lake Arendsee, NE Germany: first results of a geochemical monitoring during the restoration experiment. Sci Total Environ 218:33–44
41. Sø HU, Postma D, Jakobsen R, Larsen F (2011) Sorption of phosphate onto calcite; results from batch experiments and surface complexation modeling. Geochim Cosmochim Acta 75(10):2911–2923
42. Tsai W-T (2013) Microstructural characterization of calcite-based powder materials prepared by planetary ball milling. Materials 6:3361–3372. https://doi.org/10.3390/ma6083361
43. Walpersdorf E, Neumann T, Stüben D (2004) Efficiency of natural calcite precipitation compared to lake marl application used for water quality improvement in an eutrophic lake. Appl Geochem 19:1687–1698. https://doi.org/10.1016/j.apgeochem.2004.04.007
44. Wauer G, Teien H-C (2010) Risk of acute toxicity forfish during aluminium application to hardwater lakes. Sci Total Environ 408:4020–4025. https://doi.org/10.1016/j.scitotenv.2010.05.033

45. Weismantel GE, Armstead JC (1999) Whiting (calcium carbonate). In: McKetta JJ (ed) Encyclopedia of chemical processing and design: volume 67—water and wastewater treatment: protective coating systems to zeolite. CRC Press, Boca Raton
46. Welch EB, Gibbons HL, Brattebo SK, Corson-Rikert HA (2017) Distribution of aluminium and phosphorus fractions following alum treatments in a large shallow lake. Lake Reserv Manag 33:198–204. https://doi.org/10.1080/10402381.2016.1276653
47. Xu N, Chen M, Zhou K, Wang Y, Yin H, Chen Z (2014) Retention of phosphorus on calcite and dolomite: Speciation and modeling. RSC Advances 4(66):3525–35214. https://doi.org/10.1039/C4RA05461J
48. Yavus Ö, Guzel R, Aydin F, Tegin I, Ziyadanogullari R (2007) Removal of cadmium and lead from aqueous solution by calcite. Pol J Environ Stud 16(3):467–471
49. Zamparas M, Kyriakopoulos GL, Drosos M, Kapsalis VC, Kalavrouziotis IK (2020) Novel composite materials for lake restoration: a new approach impacting on ecology and circular economy. Sustainability 12(8):1–17. https://doi.org/10.3390/su12083397
50. Zhang Y, Prepas EE (1996) Short-term effects of $Ca(OH)_2$ additions on phytoplankton biomass: a comparison of laboratory and in situ experiments. Water Res 30(5):1285–1294. https://doi.org/10.1016/0043-1354(95)00273-1

Chapter 7
Novel Composite Materials as P-Adsorption Agents and Their Potential Applications as Fertilizers

Chen Zhang, Xiaolei Sun, Meng Li, Miltiadis G. Zamparas, and Marios Drosos

Abstract In a world with high demand in P for the maintenance of soil fertility and crop productivity, this chapter introduced the types and sources of P fertilizers in soil. Specifically chemical fertilizers were discussed, with phosphate rock being the primary source of phosphate, versus organic phosphate fertilizers, such as P humates and organominerals. Moreover, novel composite materials that have been used as P-adsorption agents are proposed as an alternative source of soil P fertilizers, which could be effective soil conditioners, slowly releasing macro- and micro-nutrients in agricultural soils.

Keywords Phosphate · Fertilizers · Lake restoration · Humates · Bephos™

7.1 Phosphorus Fertilizers in Soil

Phosphorus (P) is one of the critical macronutrients and the deficiency of P may limit plant growth. It is the key component of some building block of life such as DNA, ATP and lipids [1]. Therefore, P plays an important role in all major metabolic processes of plants including photosynthesis, respiration and energy transfer [2]. Plants mainly uptake P from soils through roots. Though P is stored in a large amount in nature but only a little part can be used by plants and thus it is considered as one of the

C. Zhang · M. Drosos (✉)
Institute of Resource, Ecosystem, and Environment of Agriculture (IREEA), Nanjing Agricultural University, 1 Weigang Road, Nanjing 210095, China
e-mail: drososmarios@njau.edu.cn

X. Sun
Institute of Bio- and Geosciences, Agrosphere (IBG-3), Forschungszentrum, 52425 Jülich, Germany

M. Li
Institute of Wetland Research, Chinese Academy of Forestry, Beijing 100091, China

M. G. Zamparas
Laboratory of Sustainable Waste Technology Management, School of Science and Technology, Hellenic Open University, Parodos Aristotelous 18, 26335 Patras, Greece

M. G. Zamparas and G. L. Kyriakopoulos (eds.), *Chemical Lake Restoration*,
https://doi.org/10.1007/978-3-030-76380-0_7

least biologically available nutrients [3]. This is related to the fact that the forms of P in soils are usually unavailable for plants while only the soluble inorganic form (orthophosphates) can be dissolved in soil solution and can be then absorbed by plants. P in soils is usually categorized in two forms which are the organic P and the inorganic P forms [4]. Soil organic P share a large proportion of soil P pool ranges from 30 to 80% in the agro-ecosystems [5]. Organic P is identified as a mixture of monoesters (Mono-P), diesters (Di-P) phosphates, phosphonates (Phos-P) and organic polyphosphates [6, 7]. All these forms should be hydrolysed by specific catalytic enzymes to become biologically available as a form of orthophosphates. Soil inorganic P forms can also be divided into two parts: the insoluble and the soluble forms. P can form easily insoluble complexes in soils in combination with Al, Fe and Ca. In such a case, the soluble P forms in soils may decrease significantly affecting crop productivity. This is mainly the case in soils with a large content of minerals like hematite, goethite and gibbsite that can bond P easier and reduce the available P. Therefore, external P input to the soil is necessary most of the times to guarantee crop production efficiency [8].

Fertilizer use date back nearly to the third millennium B.C. [9]. Ancient farmers used livestock manure, excrements, guano and bones as fertilizers to increase the macronutrients (N, K and P) in soils even they had no knowledge of their scientific properties. The new era for P fertilizer industry begun from the nineteenth century with the invention of organophosphates [3]. In 1840, Sir Joseph Henry Gilbert and Sir John Bennet Lawes conducted long-term field experiments to identify the effectiveness of mineral and organic fertilizers on crop yields in England [10]. In the experiment, superphosphate was applied. At that time, Baron Justus von Liebig recognized the idea that the soluble forms of P could be available for plants [3]. Sulfuric acid was used to transform other forms of P to superphosphate, making phosphate rock available as fertilizer. The fully acidulated product used by modern industry as P fertilizer includes mainly single superphosphate (SSP), triple superphosphate (TSP), diammonium phosphate (MAP), and diammonium phosphate (DAP) [9] (Table 7.1). These fertilizers are easily soluble in soils which can provide relatively more available P for plant uptake. On the other hand, partially acidulated phosphate rock fertilizers or compacted fertilizers (the combination of non-acidulated phosphate rock and water-soluble P) also occupy some market share in P industry [11].

As a major evaluation index for P fertilizer quality, solubility can be divided into three categories: water-soluble P, NAC-soluble P (neutral ammonium citrate soluble P), and citrate-soluble P [13]. However, the P in commercial-grade acidulated P fertilizer is not fully water-soluble, since usually a 10–15% is remaining water-insoluble [11]. The remaining water-insoluble P in soil may not be harmful for the crop growth but high water-soluble P forms do not necessarily indicate a high P availability in soils.

The orthophosphate form may be adsorbed by Fe–Al oxides in acidic soils and may be converted to water-insoluble P by $CaCO_3$ in calcareous soils [14, 15]. On the other hand, water-insoluble P fertilizers or partially water-soluble P fertilizers contain less water-soluble P fractions which can be agronomically effective under specific soil management conditions such as in acidic soils rich in organic matter, with high clay

Table 7.1 Comparison between different P fertilizer [12]

Fertilizer products	Countries	Mainly forms	Quality[a]		Cost (Yuan ton^{-1})
			Ci (%)	Ri (t/t of P_2O_5)	
Triple straight phosphate (TSP)	China	$Ca(H_2PO_4)_2–2H_2O$	46	1.041	815
Triple straight phosphate (TSP)	USA	$Ca(H_2PO_4)_2–2H_2O$	46	1.041	–
Monoammonium phosphate (MAP)	China	$NH_4H_2PO_4$	44	1.188	1,130
Monoammonium phosphate (MAP)	USA	$NH_4H_2PO_4$	51	1.100	–
Diammonium phosphate (DAP)	China	$(NH_4)_2HPO_4$	46	1.124	1,311
Diammonium phosphate (DAP)	USA	$(NH_4)_2HPO_4$	46	1.089	–
Nitrate phosphate (NP)	China	–	12	1.125	717
Nitrate phosphate (NP)	USA	–	12	1.125	–
Compound fertilizer (NPK)	China	–	15	1.126	1217
Compound fertilizer (NPK)	USA	–	15	1.126	–
Single superphosphate (SSP)	China	$Ca(H_2PO_4)_2–2H_2O$	14	1.157	250
Single superphosphate (SSP)	USA	$Ca(H_2PO_4)_2–2H_2O$	18–20	1.024	–
Fused calcium magnesium phosphate (FMP)	China	$Ca_3(PO_4)_2$	14	1.157	250
Fused calcium magnesium phosphate (FMP)	USA	$Ca_3(PO_4)_2$	18–20	1.024	–

[a]Ci and Ri refer to the product's quality indicated by the P_2O_5 content in the products (%) and the refined phosphate rock consumption per unit of product (t/t of P_2O_5)

content and P adsorption capacity [16]. That being so, the use of P fertilizer should match the soil type and the soil management to enhance the agronomic efficiency (Table 7.2).

Global P fertilizer usage amount was nearly 21 million tons in 2015 and about 22–27 million tons of P per year were estimated required to cover the croplands needs [20]. The P demand will remain high due to the population increase and the

Table 7.2 Available P levels, legacy soil P, P fertilizer requirement and actual P fertilizer use in some places

Study area	Land use	Soil type	Legacy soil P (kg ha^{-1})	Available P levels (mg kg^{-1})	Recommended P fertilizer application (in kg P ha^{-1})	References
Belgium (Wallonia)	Cropland	Calcareous clay soil	–	23	83	[17]
Belgium (Wallonia)	Cropland	Loamy acid soil	–	21.7	–	[17]
Canada (Alberta)	Cropland	–	–	100	0	[18]
Canada (Manitoba)	Cropland	–	–	20	34	[18]
Canada (Prince Edward Island)	Cropland	–	–	416	100	[18]
China (Heilongjiang)	Cropland	–	223.5	–	34.6	[19]
China (Jilin)	Cropland	–	226.5	–	48.2	[19]
China (Liaoning)	Cropland	–	461.6	–	52.6	[19]
China (Fujian)	Cropland	–	1577.4	–	111.9	[19]
China (Guangdong)	Cropland	–	1622.3	–	103.1	[19]
China (Guangxi)	Cropland	–	1776.8	–	127.0	[19]
China (Hainan)	Cropland	–	2009.4	–	141.2	[19]
China (Beijing)	Cropland	–	1096.0	–	93.8	[19]
China (Hebei)	Cropland	–	812.4	–	74.8	[19]
China (Henan)	Cropland	–	1443.6	–	140.0	[19]
China (Shandong)	Cropland	–	1157.9	–	96.5	[19]
China (Shanxi)	Cropland	–	595.9	–	47.7	[19]
China (Tianjin)	Cropland	–	574.9	–	86.0	[19]
China (Gansu)	Cropland	–	451.9	–	32.1	[19]
China (Inner Mongolia)	Cropland	–	444.1	–	44.6	[19]
China (Ningxia)	Cropland	–	425.0	–	44.4	[19]
China (Qinghai)	Cropland	–	363.8	–	25.3	[19]
China (Shaanxi)	Cropland	–	715.3	–	88.4	[19]

(continued)

Table 7.2 (continued)

Study area	Land use	Soil type	Legacy soil P (kg ha^{-1})	Available P levels (mg kg^{-1})	Recommended P fertilizer application (in kg P ha^{-1})	References
China (Xinjiang)	Cropland	–	818.2	–	78.1	[19]
China (Guizhou)	Cropland	–	233.1	–	47.5	[19]
China (Sichuan)	Cropland	–	916.7	–	106.5	[19]
China (Tibet)	Cropland	–	331.2	–	46.1	[19]
China (Yunnan)	Cropland	–	875.4	–	97.1	[19]
China (Chongqing)	Cropland	–	480.5	–	65.8	[19]
China (Anhui)	Cropland	–	1337.5	–	122.2	[19]
China (Hubei)	Cropland	–	1737.0	–	153.5	[19]
China (Hunan)	Cropland	–	729.8	–	79.9	[19]
China (Jiangsu)	Cropland	–	918.4	–	88.3	[19]
China (Jiangxi)	Cropland	–	812.9	–	82.1	[19]
China (Shanghai)	Cropland	–	591.4	–	68.4	[19]
China (Zhejiang)	Cropland	–	312.4	–	58.0	[19]
Denmark	Cropland	Calcareous clay soil	–	25	–	[17]
Denmark	Cropland	Loamy acid soil	–	13	~35	[17]
England and Wales	Cropland	Calcareous clay soil		27	–	[17]
England and Wales	Cropland	Loamy acid soil	–	11.4	~35	[17]
France	Cropland	Calcareous clay soil	–	30	–	[17]
France	Cropland	Loamy acid soil	–	17	~35	[17]
Italy (Lombardy)	Cropland	Calcareous clay soil	–	59	0	[17]
Italy (Lombardy)	Cropland	Loamy acid soil	–	40	0	[17]

(continued)

Table 7.2 (continued)

Study area	Land use	Soil type	Legacy soil P (kg ha^{-1})	Available P levels (mg kg^{-1})	Recommended P fertilizer application (in kg P ha^{-1})	References
Spain (Navarre)	Cropland	Calcareous clay soil	–	25	–	[17]
Spain (Navarre)	Cropland	Loamy acid soil	–	12	–	[17]
Switzerland	Cropland	Loamy acid soil	–	28.6	~35	[17]
Hungary	Cropland	Calcareous clay soil	–	103	0	[17]
Hungary	Cropland	Loamy acid soil	–	62	0	[17]
The Netherlands	Cropland	Calcareous clay soil	–	81	–	[17]
The Netherlands	Cropland	Loamy acid soil	–	44	89	[17]
Germany	Cropland	Calcareous clay soil	–	70	–	[17]
Germany	Cropland	Loamy acid soil	–	17.5	–	[17]
Poland	Cropland	Calcareous clay soil	–	31	–	[17]
Poland	Cropland	Loamy acid soil	–	14.8	~35	[17]
USA (Florida)	Cropland	–	–	30	0	[18]
USA (Maine)	Cropland	–	–	25	0	[18]
USA (New Jersey)	Cropland	–	–	76	34	[18]
USA (New York)	Cropland	–	–	76	34	[18]
USA (Michigan)	Cropland	–	–	120	0	[18]
USA (Minnesota)	Cropland	–	–	51	84	[18]
USA (North Dakota)	Cropland	–	–	16	0	[18]
USA (Wisconsin)	Cropland	Sandy soil	–	120	34	[18]
USA (Wisconsin)	Cropland	Non-sandy soil	–	200	56–73	[18]

(continued)

Table 7.2 (continued)

Study area	Land use	Soil type	Legacy soil P (kg ha^{-1})	Available P levels (mg kg^{-1})	Recommended P fertilizer application (in kg P ha^{-1})	References
USA (California)	Cropland	–	–	25	–	[18]
USA (Colorado)	Cropland	–	–	22	67	[18]
USA (Idaho)	Cropland	–	–	20–30	0	[18]
USA (Manitoba)	Cropland	–	–	20	34	[18]
USA (Oregon Willamette valley)	Cropland	–	–	20	90	[18]
USA (Oregon Columbia basin)	Cropland	–	–	10	170	[18]
USA (Oregon other areas)	Cropland	–	–	40	90	[18]
USA (Texas)	Cropland	–	–	60	0	[18]
USA (Washington)	Cropland	–	–	20	0	[18]

expansion of biofuel industry [21]. High external P inputs are not always effective to increase crop yields, since the P fertilizer usage efficiency is low. It is believed that only 10–15% of P could be recovered by crops yearly [22]. The part of P which can't be used by plants could be sequestered in the soil with time in non-bioavailable forms. The accumulation of P in soils is named 'Legacy P' [23]. Legacy P is not irreversibly fixed in soils and it can become available to crops under different conditions [24]. Except the P fixed by oxides or $CaCO_3$ in soils, leaching and surface runoff loss is another pathway for residual P in soils.

P losses from soil to waterbody may cause an adverse effect on the terrestrial environment [25, 26]. P is considered as the limited nutrient for alga bloom in surface water because some species of alga such as *Cyanobacteria* can biologically deposit N from the atmosphere. P loss through runoff from cropland is a primary contributor to the eutrophication of surface waters [27, 28]. And the fact that significant loss quantities of P were detected in leachates by soil column or field lysimeters revealed that P leaching also poses a great risk to the environment [29, 30]. Many researches have been conducted to predict potential P release from soils via laboratory analysis of soil P test [31]. It is suggested that both agronomic and environmental indexes (e.g. Mehlich-3-P, Olsen P and Bray-P, 0.01 mol L^{-1} $CaCl_2$-extractable P [$CaCl_2$-P], and Fe-oxide coated filter paper strip extractable P [FeO-P]) can be used to estimate P loss [32, 33]. Specifically, when $CaCl_2$-P concentrations are dramatically increased, the 'change point' which is the special point between soil Olsen-P and $CaCl_2$-P was used to predict the P loss through leaching [34, 35].

Besides the assessment of P loads from soils to the water body, P fertilizer management is another study hotspot to control P loss. P loss from arable lands to waterbody can be affected by many factors such as the type of P fertilizers, the application of P amendments, the irrigation management, the soil characteristics and the rainfall regimes. The use of water-soluble P fertilizers is one of the predominant factors inducing eutrophication [36]. Wen et al. [37] found that the combination of manure and chemical fertilizers can increase maize yield and nutrient uptake in a long-term field experiment. The study of Pizzeghello et al. [32] confirmed that the type of soil, the treatment and the depth affected the P sorption in soils and also showed the potential of manure on controlling P loss. Moreover, the research of Liu et al. [38] identified the effect of rainfall regimes and fertilizer application method on the P losses via surface runoff. Considering the adverse effect of the excessive use of P fertilizer on the environment, measures on improving P use efficiency have been proposed including the modification of the surface soil properties, the changing of P fertilizer type, the optimization of P fertilizer application placement, the rate and timing of the application, the managing irrigation models and the appropriate soil amendments use [39, 40].

Creating P fertilizer products based on new technologies and understanding the plant physiological processes is regarded as the novel way to improve plant uptake of P [41]. For example, nanotechnology can be used to improve P fertilizer use efficiency because nano-sized particles can move easily in the soil and be captured by the plant roots [42, 43]. Montalvo et al. [44] found that hydroxyapatite nanoparticles showed a potential to increase the P uptake and the percentage of P in plants. Samavini et al. [45] certified that nano-hydroxyapatite can enhance root growth and improve P availability. Recycling P to lower new P input is also an innovative method to decrease environmental risk. Many researchers use struvite (ammonium-magnesium phosphate) to achieve P recycling from sludge and wastewater [46]. The recycling of agricultural by-products to produce organomineral fertilizers which can work as soil amendments to improve soil physical and chemical characteristics is also a feasible choice. Herzel et al. [47] used sugarcane byproducts to produce organomineral phosphate fertilizer which was subsequently applied in crops to increase sugarcane yield. Sewage sludge ash was also promoted to be a promising secondary P source for fertilizer production when suitable purification methods were applied [47]. Some studies proved that the application of micronutrients can contribute to native P mobilization resulting in the improvement of P use efficiency without additional P input [41]. The importance of micronutrients such as zinc and iron in stimulating P availability under low soil P content has been reported [48, 49], raising the concept of beneficial nutrient interactions in nanoscale to improve P use efficiency in soil.

Accumulation of hazardous trace elements such as arsenic (As), cadmium (Cd) and lead (Pb) in croplands is another environmental problem caused by excessive P fertilizer application [50]. As a primary source for phosphate fertilizer, sedimentary phosphate rock may include heavy metals and other toxic elements. Though the manufacturing processes technologies for P fertilizers are improved now, adulterated products with toxic elements cannot be avoided. Enrichment of toxic trace elements may not only result in soil degradation but also pose serious threats to human health through their migration into the food chain. On the other hand, removal of trace elements added by fertilizer usually depends on crop harvesting [51]. Fertilization can improve crop yield and enhance trace elements removal from soil. Therefore, to find a proper soil remediation strategy taking into account the effect of fertilizer on trace elements accumulation is an extremely complicated subject.

Cd is one of the prime trace elements accumulated in soils with the application of P fertilizer. Cd is toxic and even low-level exposure to it may result in diabetes, hypertension and cancer [52]. Phosphorus-ore with cadmium content is the main source for manufacturing P fertilizer, though the concentration of Cd content bonded with P in apatite can vary widely. The Cd concentration of rock phosphates from various countries around the world ranged from 0.2 mg Cd/kg phosphate rock to 100 mg Cd/kg phosphate rock [11]. Even after the strong acidulation during the P fertilizer formation process with H_2SO_4, all the Cd content can remain in SSP [53]. Cd is not only present in chemical fertilizers but also in organic fertilizers such as pig and cattle manure [54]. The research of tracing freshly applied Cd through mineral P fertilizers showed that the Cd recoveries were comparatively low in soil-fertilizer-plant systems, which result in the build-up of a residual Cd pool in soils [55, 56]. The Cd availability is affected by soil characteristics including soil pH, organic matter content and components of solid-phase mineral, similar to those affecting the P availability of Ca–P [11, 50]. It's the precipitation and co-precipitation process that determines the Cd mobility in alkaline soil. Some studies revealed that water-soluble P compounds may change mobile Cd into less mobile forms through adsorption onto the solid phase to reduce the Cd uptake [57]. This process is also mainly influenced by soil pH.

Pb is a toxic heavy metal introduced in soils by the application of P fertilizer. It can be strongly fixed in the soil through a series of chemical reactions like adsorption, ionic exchange and precipitation [58]. The content of clay minerals, organic matter and iron, manganese and aluminum hydroxides in soil are the adsorption agents for Pb-compounds [59]. Pb compounds in soil are mostly insoluble and easy to accumulate since they hardly undergo microbial or chemical degradation [60]. However, Pb complexes solubility can be changed when soil pH fluctuates [61]. Pb-organic complexes can form in alkaline soil [59]. Fine soils with high organic matter content contain more Pb than soils with coarse texture and low organic matter content [50]. Though mineral P fertilizer is considered as one of the main sources of Pb input to soil, numerous researchers have identified that phosphate-based materials can reduce the leachability of Pb and decrease the availability for plants [62].

Nearly 90% of As input with fertilizer application was detected to leach below root zone over time [63]. As a carcinogen, arsenic can pose a severe threat to animal and human health [64]. Though As is a nonmetallic element, it shows similar chemical characteristics with metals and researchers usually regarded As as a metal [65]. Arsenic accumulation in soils is not only due to fertilizer application but also due to atmospheric deposition [66]. Some studies even certificated that P fertilizer can't cause a serious As accumulation in soil [58]. Woolson et al. [67] found that As content would increase with the fertilizer application in soils with high iron content.

In soils poor in iron, As can be immobilized by forming complexes with aluminum or exchangeable calcium [50].

7.2 Phosphate Rock as a Primary Source for Phosphate Fertilizer

Phosphate rock (PR) is the primary stock source for P fertilizer production. After world war II, the use of mineral P sources raised extremely and at that time, phosphate rock was regarded as an abundant and cheap P source [68]. P from mined phosphate rock is indispensable for modern societies. The use of P includes a variety of industries' products, such as food and feed additives, detergents and others. But the dominant driving force for world PR depletion was considered food production [21]. From 1961 to 2013, the total consumption has increased fivefold and was up to 31 million metric tons [69]. According to summary of available data, from the 290 Gt PR which are reserved in earth, 60 Gt could be extracted for use [70]. However, 95% of the world's PR is used for manufacturing P fertilizer [71].

In fact, it is usually difficult to assess the geological stock of PR in the world because not all the information was reported and only the part of the reserve which could be extracted and mined is counted [72] (Fig. 7.1). Though is difficult to identify the geological stock of PR accurately and its depletion timelines, it is sure that PR is a non-homogenous and non-renewable resource. Some studies estimate approximately 50–100 years of current known reserves still remain [21, 73]. On the other hand, the costs for mining new PR will be higher in the future with the

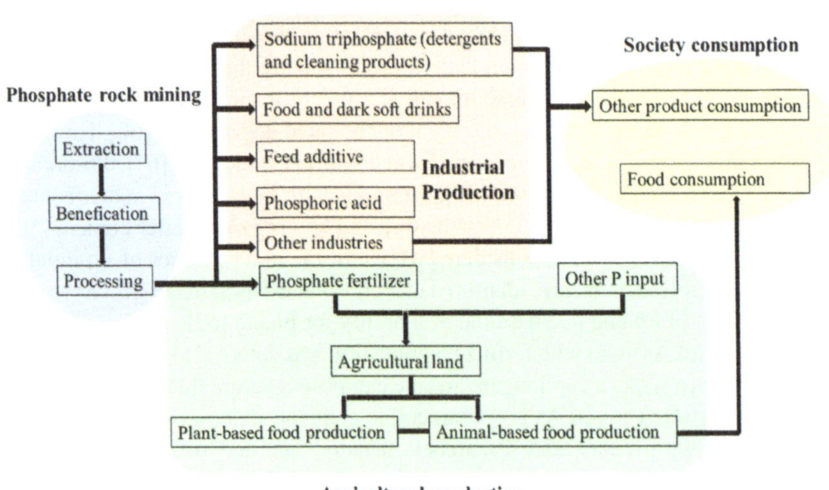

Fig. 7.1 Phosphate rock flow analysis [69]

higher quality PR reserves depleted [74, 75]. Considering the urgent situation to decrease PR mining, improving the phosphorus use efficiency (PUE) becomes necessary in industry and agriculture [76]. Only 22% of the extracted PR is being used for promoting human food which is really needed in the agricultural systems [69]. Researchers have suggested many methods to increase PUE through the management of plant, soil, fertilizer and mining [77]. The measures were detailed as P substitution, detergent reduction, food wastage reduction, food waste recycling, urban wastewater treatment, bio-fertilizer application, no-till agricultural expansion and phosphorus wetland absorption [78].

Using secondary P in available bioresources (animal manure and sewage sludge) is a popular choice to substitute the traditional fertilizers, which is beneficial to reduce the dependency on PR [79]. P in manure and sewage sludge was certificated to be more recyclable than insoluble inorganic fertilizer [80]. However, secondary P fertilizer usage efficiency is also limited by some factors including geographic separation between the place of generation and croplands application [81]. Many studies have posed to identify secondary P fertilizer potential and solve the limited factors for efficiency usage.

PR is referred to rock with the apatite group. It occurs in three primary environments including (a) fluorapatite, mainly contained in igneous and metamorphic environments, (b) hydroxyapatite, in igneous, metamorphic environments and biogenic deposits, (c) carbonate hydroxyapatites, in inlands and caves [82, 83]. More than 40 countries all over the world are mining PR. The main countries for P fertilizer production are USA, Morocco and China. USA is the world's largest producer and consumer of PR [63]. The source of PR and acidulation method for fertilizer manufacturing are the main reasons for the differences in the water-insoluble P fraction chemical composition [11]. PR forms and sources are also the reason for the various trace elements detected [83, 84]. PR usually contains higher radionuclide concentrations than in the natural levels. Therefore PR mining process could redistribute dangerous radionuclides in soil and plants accompanied by P fertilizer application [85].

7.3 The Role of Humics in Organic Phosphorous Availability

Humus was named firstly by De Saussure [86] taken from the Latin word for earth-ground, which originated from the Greek word 'ΧΩΜΑ'. Humus denominates the fraction of soil organic matter that consists of decomposed plant and animal matter by microbes and oxidation. The main components of soil humus are humic substances. Achard was the first scientist that isolated and fractionated humic substances from peat [87].

Traditionally humic substance understanding was based on the acid–base theory that classified the organic acids of soil organic matter (SOM) as humic/fulvic acids and their conjugated bases as humates. So, humic acids are extracted from soil when a strong-base extract is acidified, while fulvic acids are the organic acids that do not precipitate after acidification. The insoluble soil organic residue is called humin.

The transformation of the structural components in biotic cells of soil to the amorphous humic state is called humification.

There were three main pathways proposed to explain humification. The lignin theory of Waksman [88], the polyphenol theory of Flaig [89] and the sugar-amine condensation theory of Maillard [90]. However, these theories alone aimed to explain the early understanding of humification as a recombination of degraded biopolymers to new macromolecular structures. A recent understanding views humic substances not as macropolymers but as heterogeneous and relatively small in size molecular components of SOM that are auto-assembled in meta-stable supramolecular associations, held together by π–π stacking, hydrogen bonds and metal bridges [91, 92]. Actually their ability to stabilize free radicals [93] and their environmental recalcitrance [94] is an outcome of their meta-stability. They constantly evolve according to the factors that tend to destroy their assembly, to a new recombined form which incorporates the invader factors into their main body, precisely following Le Chatelier's principle.

So, humification is the dynamic process of weakly bound superstructures that increase their proximities, gradually raising the hydrophobicity of the system. Since more humified material is more hydrophobic, it is less accessible to microbial degradation and mineralization. This humic pool can be hydrophobically adsorbed on clay and chemically sorbed by oxides and can be again released to the soil solution due to plant exudation. This solid to liquid equilibrium of humic molecules, which play a bioactive role in plant growth, is responsible for the physicochemical fertility of soil [95].

Humic substances are composed of C, H, O, N, P, S and a small amount of Ca, Mg, Fe, Si and other ash elements [96]. Different from other soil compounds, humic substances reveal specific structural characteristics and their chemical composition is considered to represent natural organic matter (NOM) [97]. Their aggregation degree can be affected by pH, ionic strength and mineral composition of the solution [98]. The presence of polar groups in humics affect their solubility and molecular size [99]. Many analytical techniques have been used to study HS including elemental analysis, proton-binding, UV-Visible spectrophotometry, FTIR and solid-state NMR [100, 101].

Humic substances account for nearly 25% of the total organic carbon on the earth and are widely spread in both soil, water, lake sediments, peat, brown coal and shales [102]. Therefore, humic substances acquire important ecological and environmental functions [103]. They maintain the soil fertility, improve the soil quality, sustain plant growth and enhance environmental health. In fact, humic substances can influence the soil structure and affect the chelation of mineral elements. Humic substances have also the potential to improve plant growth, but their effectiveness depends on their source, the climatic conditions, the soil pH and the type of plants being treated [104]. In fact, some researchers regarded humic substances as biostimulants in horticulture based on their effects on nutrient uptake and plant metabolism [105]. Humic substances can alter both structural and physiological changes in roots and shoots, which result in changes in nutrient uptake, assimilation and distribution [106]. They can also justify the abiotic stress tolerance by shifting the primary and secondary metabolism of plants [105]. The metal complexation ability of humics can improve plant iron nutrition, which is in accordance with the idea that humic substances may induce changes in root morphology and plant metabolism [107]. On the other hand, humics can regulate microbial growth in soil. Wei et al. [108] identified that HA suppressed phytopathogenic fungi growth. So, the study of humic substances can provide useful information about the metal ion behavior in soil and water, the soil N cycle and SOM microbial interactions (Fig. 7.2), and can guide to environmental-friendly strategies adaptation to mitigate pollution and enhance plant growth.

Considering that humic substances can be depleted from soils due to wind and water erosion and leaching, their replenishment is important. Therefore, many humic products are applied in agricultural fields to increase crop yield and maintain soil quality [109]. These humic products are mainly concentrated aqueous solutions of potassium or sodium humates, which are produced by alkali extraction from organic matter like lignite, leonardite or peat [110]. Similar to humic substances, humates are also recognized to be helpful in modifying biological properties of soil, enhancing nutrition efficiency, abiotic stress tolerance and crop quality traits [105, 111, 112]. Since humic substances have the potential to accumulate macro- and micro-nutrients into their structure, the future in agrochemistry can be the application of humic materials high in phosphorus content, which can act as slow-release fertilizers. Actually,

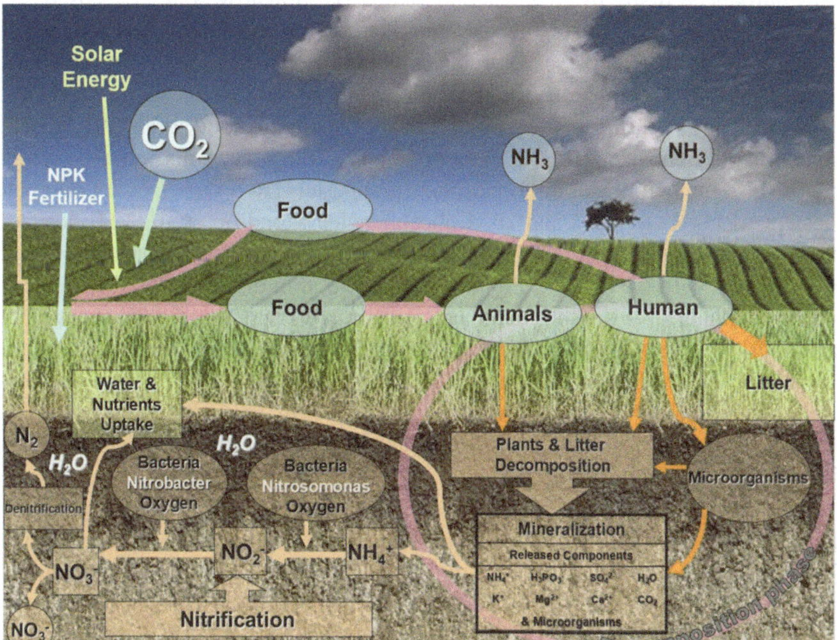

Fig. 7.2 Biological cycle of soil organic matter

phosphorous is ubiquitous in natural organic matter (NOM) deriving from phospholipids, nucleotides and metabolites [113]. As a matter of fact, 85% of the soil P is found in organically bound forms, while the content of P in soil organic matter ranges from >1% up to 3% [114]. John et al. [115] reported organic C/P ratio ranging from 46 to 648, and it has been found that this ratio increases with the increase of soil aggregate [116] and nominal humic substances [117] size.

Nucleic acids P in soil occurs only at 0–3% of the total organic P (OP), being the rarest form, while inositol phosphates (phytin or phytic acid) occur up to 60%, being the most abundant form. Phytin, phospholipids and nucleic acids in soil are mainly deriving from plants, however it seems that soil is a better pool for OP forms that they are present in low concentrations in plants. For example, phytin, which derives mainly from beans and leafy vegetables, accumulates in high concentrations in soil due to its tendency to form complexes with Fe, Al and Ca, and when hydrolyzes it releases phosphate groups [118, 119]. However, besides myo-inositol deriving in soil from plant decay, scyllo-inositol is also present, deriving from in-situ synthesis by microorganisms. Phospholipids are present in soil OP up to 7% and they derive from microbes. Nucleic acids, containing phosphoric acid residue, can be found in soil up to 3%, while phosphate metabolites are present in smaller amounts. Nevertheless, 30–50% of the soil OP is hidden inside humic substances, masked and still unidentified, even though it is assumed that it is usually present in the form of simple phosphate esters [114]. Nanny and Minear [120] found mono- and di-ester phosphates in lake water NOM. ^{31}P-NMR has been used in studies of alkaline extracts of SOM and soils introduced firstly by Newman and Tate [121]. Makarov et al. [122] used ^{31}P-NMR to research phosphate–metal–humic interactions, revealing the formation of phosphorus–humic complexes. Norman et al. [123] found that when pyrophosphate is used for humic extraction it is incorporated as a contaminant inside the humic structure. This is a further evidence of the humic–P complexation affinity. Besides the inorganic orthophosphates (PO_4^{3-}) found in humic substances, phosphorous can be present also in the forms of orthophosphate monoesters ($ROPO_3^{2-}$), orthophosphate diesters ($(RO)(R'O)PO_2^{-}$) and alkyl phosphonates. Condron et al. [124] detected phosphonates, pyrophosphate, polyphosphonates and orthophosphates in alkaline SOM extract with the use of ^{31}P-NMR as well. So, liquid state ^{31}P-NMR is able to identify OP and distinguish its various forms in soils and humic substances [125, 126]. Various researchers identified orthophosphate monoesters as the main organic form of SOM and humic substances [127, 128]. ^{31}P-NMR spectroscopy was also used to detect humification of sewage sludge during its anaerobic treatment [129]. According to this work, pyrophosphates decrease while orthophosphates increase following the humification process.

It has been reported that fulvic acid contains 5 times more OP than humic acid. This might be due to the acidification step during the humic/fulvic separation since fulvic acid is the hydrosoluble form in low pH that can bind the released cations, which will consequently bind OP. As a matter of fact, it is usual that a pair of adjacent Fe^{3+} cations can form binuclear surface complexes with the HPO_4^{2-} oxyanion in soil, containing iron oxides such as goethite. So, there is strong affinity between Fe and P in soil [130]. Therefore Fe-binding in humic substances can be the platform to create phosphorus–iron-humates (Fig. 7.3).

In fact, dissolved organic matter (DOM) adsorbed on mineral surfaces can displace P previously attached to Fe and Al hydroxides, increasing its availability in soil [92]. Moreover, humic substances can also interact with phosphate anions forming metal bridges and enhance P solubilisation and plant uptake. So, humic matrix can pile up OP from plant and animal decay and provide it slowly to the plants forced by microbial decomposition and root exudation (Fig. 7.4).

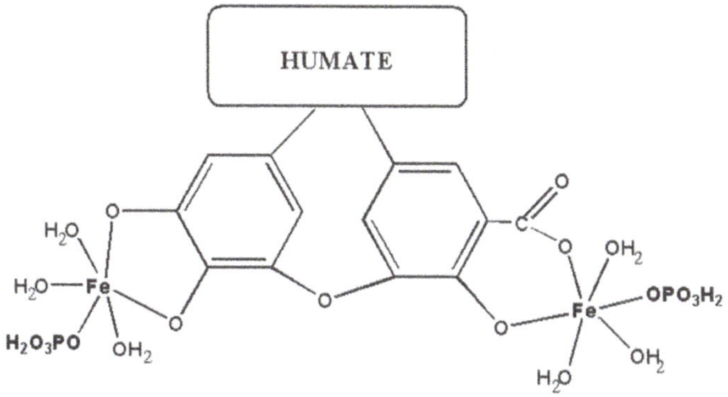

Fig. 7.3 FeP binding in humic substances

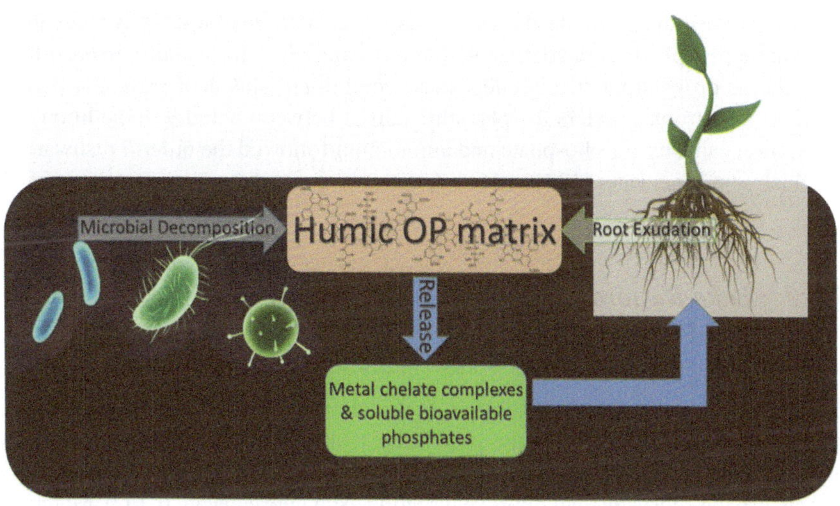

Fig. 7.4 Phosphorous release mechanism from SOM

7.4 Potential Usefulness of Lake Restoration Materials as Soil Improvers and Fertilizers

7.4.1 The Novel Bentonite Humic Acid Composite Material Bephos™

Bephos™ (nomination is originated from the words 'Bentonite' and 'phosphorus') is a novel low-cost composite material embedding Fe, Cu ions and humic acid in the interlayer space of a natural bentonite. This material is derived from a novel precursor Fe-bentonite [131, 132]. The point of zero charge (PZC) of the material is pH 8.8. This implies that the positive charges of bound Fe^{3+}, Cu^{2+} and the clay surface groups counterbalance the negative charges of the Humic acid. From the functional point of view, the existence of both positively charged sites, e.g. Fe^{3+}, Cu^{2+}, as well as negatively charged sites (e.g. due to HA) renders Bephos™ the capability to bind both anionic species, e.g. PO_4, as well cationic species, e.g. $NH^{+}{}_{4}$. In addition, the important information from XRD data [133] is that Fe and Cu ions have been incorporated in solid crystalline phases, while other phases have been depleted from the pristine clay. Humic acid is rich in carboxylates which act as strong anchoring sites for Fe and Cu ions. Thus, the formed COO–Fe, COO–Cu sites behave as coordination sites for P and N due to the known strong affinity of Fe and Cu to coordinate P and N atoms. Bephos™ is a novel composite material where Fe, Cu, HA and clay lamellas are strongly interconnected [134].

Moreover, the modified clay was highly efficient for simultaneous phosphorus removal from water at the wide range of pH values 5–9. Its efficiency is maintained at alkaline pH (pH 8–9) in contrast with most materials, whose ability in adsorbing phosphorus drops dramatically under these conditions. This is of particular importance as in eutrophic waters the pH value ranges between 8 and 9. In addition, the adsorption capacity for phosphate and ammonium followed the order: Freshwater > brackish > marine water [133].

7.4.2 Lake Restoration Materials as Soil Improvers and Fertilizers

In the previous chapters, P-adsorption materials have been studied only for lake restoration. However, there is the possibility to reuse/recycle these materials for agricultural use. Most of these materials are clay-based, so their addition to soil may not disturb the ecosystem but improve its cation exchange capacity (CEC). Materials like Bephos™ contain also monovalent iron and copper that could act like micronutrients [135, 136]. However, their most advantageous feature is their ability to retardedly adsorb phosphorous and, then, to slowly release it. Specifically, it has been reported that Bephos™ can uptake 26.5 g kg^{-1} phosphate, gradually releasing

5.3 g kg^{-1} every 6 months [132]. Considering the need of phosphorous in crop fields that is ~50 kg ha^{-1} per year (for the crops of wheat, maize and tomato), the need of 5 tn ha^{-1} of Bephos™ is estimated, which would retardedly release the necessary phosphorus over a period of more than 3 years. Bephos™ is not just a modified clay, but it can be considered as a modified organo-mineral clay since it contains humic acid [132]. The application of 5 tn ha^{-1} Bephos™ in the field would result in the addition of 6 kg ha^{-1} humic acid. However, the aforementioned results are obtained for Bephos™, because there is lack of information about other P-adsorption derivatives.

Moreover, the P-release ratio can be measured in water, not in soil. It is also noteworthy to note the need for lowering the cost of these materials since these are proven effectively recyclable in agriculture after being used in lake restoration. However, further research is needed to estimate the actual release of bound phosphorous of these materials in soil and their interaction mechanisms with mycorrhizal.

7.5 Conclusions

Humic–metal–phosphate complexes can decrease phosphate fixation in soils and increase plant growth and phosphate uptake. Phosphorus fertilizers, containing humic–metal–phosphate complexes are proven effective to improve plant growth and P uptake concerning conventional fertilizers, such as single superphosphate. The values of parameters related to plant phosphorus-utilization efficiency (PUt E) suggest that the regulation of root acquisition of phosphate from these complexes could introduce a controlling system for the optimization of metabolic P utilization in the shoot, and another controlling system that involves stress responses of roots under phosphorus deficiency.

The properly designed application of chemical restoration methods can reduce the concentration of nutrients and improve the water quality in eutrophic ecosystems. However, the demand for fertilizer is increasing worldwide, contrary to the global phosphate supply shortage [137]. It seemed that the application of phosphate fertilizers has reached its limits and it is necessary to be transformed sustainably [138]. Moreover, phosphate fertilizers losses not only reduce nutrients efficiency, resulting in low plant yields but also create adverse impacts on the environment. Thus, the use of P-adsorption agents [139–141] can prove alternative methods in which the adsorbed phosphorus can be used as fertilizer or enriching soil conditioners. Therefore, it is generated a sustainable cycle taking into account a renewable resource-based Phosphorus Adsorption Agent (P-AA) leading to environmental protection and conservation of natural resources. P-AA for removal and recovery of phosphorus from eutrophic water bodies is a promising technique but requires further study in its feasible scaling up to the agricultural sector worldwide.

References

1. Khan MS, Zaidi A, Ahemad M, Oves M, Wani PA (2010) Plant growth promotion by phosphate solubilizing fungi—current perspective. Arch Agron Soil Sci
2. Carstensen A, Herdean A, Schmidt SB, Sharma A, Spetea C, Pribil M, Husted S (2018) The impacts of phosphorus deficiency on the photosynthetic electron transport chain[1[OPEN]]. Plant Physiol
3. Ashley K, Cordell D, Mavinic D (2011) A brief history of phosphorus: from the philosopher's stone to nutrient recovery and reuse. Chemosphere
4. George TS, Giles CD, Menezes-Blackburn D, Condron LM, Gama-Rodrigues AC, Jaisi D, Lang F, Neal AL, Stutter MI, Almeida DS et al (2018) Organic phosphorus in the terrestrial environment: a perspective on the state of the art and future priorities. Plant Soil
5. Tiessen H (2008) Phosphorus in the global environment
6. Cade-Menun B, Liu CW (2014) Solution phosphorus-31 nuclear magnetic resonance spectroscopy of soils from 2005 to 2013: a review of sample preparation and experimental parameters. Soil Sci Soc Am J
7. Li M, Cozzolino V, Mazzei P, Drosos M, Monda H, Hu Z, Piccolo A (2018) Effects of microbial bioeffectors and P amendements on P forms in a maize cropped soil as evaluated by ^{31}P-NMR spectroscopy. Plant Soil
8. Frazão JJ, de Melo Benites V, Ribeiro JVS, Pierobon VM, Lavres J (2019) Agronomic effectiveness of a granular poultry litter-derived organomineral phosphate fertilizer in tropical soils: soil phosphorus fractionation and plant responses. Geoderma
9. Scholz RW, Roy AH, Hellums DT, Ulrich AE, Brand FS (2014) Sustainable phosphorus management: a global transdisciplinary roadmap. ISBN 9789400772502
10. Experiments L (2006) Rothamsted long-term experiments—guide to the classical and other long-term experiment, datasets and sample archive. Rothamsted Research, Harpenden
11. Chien SH, Prochnow LI, Tu S, Snyder CS (2011) Agronomic and environmental aspects of phosphate fertilizers varying in source and solubility: an update review. Nutr Cycl Agroecosystems
12. Zhang W, Ma W, Ji Y, Fan M, Oenema O, Zhang F (2008) Efficiency, economics, and environmental implications of phosphorus resource use and the fertilizer industry in China. Nutr Cycl Agroecosystems
13. Johnston AE, Richards IR (2003) Effectiveness of the water-insoluble component of triple superphosphate for yield and phosphorus uptake by plants. J Agric Sci
14. Gypser S, Hirsch F, Schleicher AM, Freese D (2018) Impact of crystalline and amorphous iron- and aluminum hydroxides on mechanisms of phosphate adsorption and desorption. J Environ Sci (China)
15. von Wandruszka R (2006) Phosphorus retention in calcareous soils and the effect of organic matter on its mobility. Geochem Trans
16. Chien SH, Menon RG (1995) Factors affecting the agronomic effectiveness of phosphate rock for direct application. Fertil Res
17. Jordan-Meille L, Rubæk GH, Ehlert PAI, Genot V, Hofman G, Goulding K, Recknagel J, Provolo G, Barraclough P (2012) An overview of fertilizer-P recommendations in Europe: soil testing, calibration and fertilizer recommendations. Soil Use Manag
18. Rosen CJ, Kelling KA, Stark JC, Porter GA (2014) Optimizing phosphorus fertilizer management in potato production. Am J Potato Res
19. Ma J, Liu Y, He W, He P, Haygarth PM, Surridge BWJ, Lei Q, Zhou W (2018) The long-term soil phosphorus balance across Chinese arable land. Soil Use Manag
20. Mogollón JM, Beusen AHW, van Grinsven HJM, Westhoek H, Bouwman AF (2018) Future agricultural phosphorus demand according to the shared socioeconomic pathways. Glob Environ Change
21. Cordell D, Drangert JO, White S (2009) The story of phosphorus: global food security and food for thought. Glob Environ Chang

22. Roberts TL, Johnston AE (2015) Phosphorus use efficiency and management in agriculture. Resour Conserv Recycl
23. Rodrigues M, Pavinato PS, Withers PJA, Teles APB, Herrera WFB (2016) Legacy phosphorus and no tillage agriculture in tropical oxisols of the Brazilian savanna. Sci Total Environ
24. Johnston AE, Poulton PR, Fixen PE, Curtin D (2014) Phosphorus: its efficient use in agriculture. In: Advances in agronomy
25. Bünemann EK (2015) Assessment of gross and net mineralization rates of soil organic phosphorus—a review. Soil Biol Biochem
26. Ceulemans T, Merckx R, Hens M, Honnay O (2011) A trait-based analysis of the role of phosphorus vs. nitrogen enrichment in plant species loss across North-West European grasslands. J Appl Ecol
27. Hua L, Liu J, Zhai L, Xi B, Zhang F, Wang H, Liu H, Chen A, Fu B (2017) Risks of phosphorus runoff losses from five Chinese paddy soils under conventional management practices. Agric Ecosyst Environ
28. Murphy PNC, Stevens RJ (2010) Lime and gypsum as source measures to decrease phosphorus loss from soils to water. Water Air Soil Pollut
29. Fortune S, Lu J, Addiscott TM, Brookes PC (2005) Assessment of phosphorus leaching losses from arable land. In: Proceedings of the plant and soil
30. Li M, Hu Z, Zhu X, Zhou G (2015) Risk of phosphorus leaching from phosphorus-enriched soils in the Dianchi catchment, Southwestern China. Environ Sci Pollut Res
31. Wang YT, Zhang TQ, O'Halloran IP, Tan CS, Hu QC, Reid DK (2012) Soil tests as risk indicators for leaching of dissolved phosphorus from agricultural soils in Ontario. Soil Sci Soc Am J
32. Pizzeghello D, Berti A, Nardi S, Morari F (2016) Relationship between soil test phosphorus and phosphorus release to solution in three soils after long-term mineral and manure application. Agric Ecosyst Environ
33. Jalali M, Jalali M (2017) Assessment risk of phosphorus leaching from calcareous soils using soil test phosphorus. Chemosphere
34. Maguire RO, Sims JT (2002) Soil testing to predict phosphorus leaching. J Environ Qual
35. Heckrath G, Brookes PC, Poulton PR, Goulding KWT (1995) Phosphorus leaching from soils containing different phosphorus concentrations in the Broadbalk experiment. J Environ Qual
36. Alexander RB, Smith RA, Schwarz GE, Boyer EW, Nolan JV, Brakebill JW (2008) Differences in phosphorus and nitrogen delivery to the Gulf of Mexico from the Mississippi River Basin. Environ Sci Technol
37. Wen Z, Shen J, Blackwell M, Li H, Zhao B, Yuan H (2016) Combined applications of nitrogen and phosphorus fertilizers with manure increase maize yield and nutrient uptake via stimulating root growth in a long-term experiment. Pedosphere
38. Liu J, Zuo Q, Zhai LM, Luo CY, Liu HB, Wang HY, Liu S, Zou GY, Ren TZ (2016) Phosphorus losses via surface runoff in rice-wheat cropping systems as impacted by rainfall regimes and fertilizer applications. J Integr Agric
39. Al-Rawajfeh AE, AlShamaileh EM, Alrbaihat MR (2019) Clean and efficient synthesis using mechanochemistry: preparation of kaolinite–KH_2PO_4 and kaolinite–$(NH_4)_2HPO_4$ complexes as slow released fertilizer. J Ind Eng Chem
40. Yan X, Wei Z, Hong Q, Lu Z, Wu J (2017) Phosphorus fractions and sorption characteristics in a subtropical paddy soil as influenced by fertilizer sources. Geoderma
41. Bindraban PS, Dimkpa CO, Pandey R (2020) Exploring phosphorus fertilizers and fertilization strategies for improved human and environmental health. Biol Fertil Soils
42. Ghormade V, Deshpande MV, Paknikar KM (2011) Perspectives for nano-biotechnology enabled protection and nutrition of plants. Biotechnol Adv
43. Gogos A, Knauer K, Bucheli TD (2012) Nanomaterials in plant protection and fertilization: current state, foreseen applications, and research priorities. J Agric Food Chem
44. Montalvo D, McLaughlin MJ, Degryse F (2015) Efficacy of hydroxyapatite nanoparticles as phosphorus fertilizer in andisols and oxisols. Soil Sci Soc Am J

45. Samavini R, Sandaruwan C, De Silva M, Priyadarshana G, Kottegoda N, Karunaratne V (2018) Effect of citric acid surface modification on solubility of hydroxyapatite nanoparticles. J Agric Food Chem
46. Ohtake H, Tsuneda S (2018) Phosphorus recovery and recycling. ISBN 9789811080319
47. Herzel H, Krüger O, Hermann L, Adam C (2016) Sewage sludge ash—a promising secondary phosphorus source for fertilizer production. Sci Total Environ
48. Raliya R, Tarafdar JC, Biswas P (2016) Enhancing the mobilization of native phosphorus in the mung bean rhizosphere using ZnO nanoparticles synthesized by soil fungi. J Agric Food Chem
49. Zahra Z, Arshad M, Rafique R, Mahmood A, Habib A, Qazi IA, Khan SA (2015) Metallic nanoparticle (TiO_2 and Fe_3O_4) application modifies rhizosphere phosphorus availability and uptake by Lactuca sativa. J Agric Food Chem
50. Jiao W, Chen W, Chang AC, Page AL (2012) Environmental risks of trace elements associated with long-term phosphate fertilizers applications: a review. Environ Pollut
51. Yu WT, Zhou H, Zhu XJ, Xu YG, Ma Q (2011) Field balances and recycling rates of micronutrients with various fertilization treatments in Northeast China. Nutr Cycl Agroecosystems
52. Satarug S, Garrett SH, Sens MA, Sens DA (2010) Cadmium, environmental exposure, and health outcomes. Environ Health Perspect
53. Dharma-wardana MWC (2018) Fertilizer usage and cadmium in soils, crops and food. Environ Geochem Health
54. Yang W, Zhuo Q, Chen Q, Chen Z (2019) Effect of iron nanoparticles on passivation of cadmium in the pig manure aerobic composting process. Sci Total Environ
55. Wiggenhauser M, Bigalke M, Imseng M, Keller A, Rehkämper M, Wilcke W, Frossard E (2019) Using isotopes to trace freshly applied cadmium through mineral phosphorus fertilization in soil-fertilizer-plant systems. Sci Total Environ
56. Spiess E (2011) Nitrogen, phosphorus and potassium balances and cycles of Swiss agriculture from 1975 to 2008. Nutr Cycl Agroecosystems
57. Siebers N, Leinweber P (2013) Bone char: a clean and renewable phosphorus fertilizer with cadmium immobilization capability. J Environ Qual
58. Wang Q, Zhang J, Xin X, Zhao B, Ma D, Zhang H (2016) The accumulation and transfer of arsenic and mercury in the soil under a long-term fertilization treatment. J Soils Sediments
59. Strawn DG, Sparks DL (2000) Effects of soil organic matter on the kinetics and mechanisms of Pb(II) sorption and desorption in soil. Soil Sci Soc Am J
60. Bolan N, Kunhikrishnan A, Thangarajan R, Kumpiene J, Park J, Makino T, Kirkham MB, Scheckel K (2014) Remediation of heavy metal(loid)s contaminated soils—to mobilize or to immobilize? J Hazard Mater
61. Zeng G, Wan J, Huang D, Hu L, Huang C, Cheng M, Xue W, Gong X, Wang R, Jiang D (2017) Precipitation, adsorption and rhizosphere effect: the mechanisms for phosphate-induced Pb immobilization in soils—a review. J Hazard Mater
62. Hafsteinsdóttir EG, Camenzuli D, Rocavert AL, Walworth J, Gore DB (2015) Chemical immobilization of metals and metalloids by phosphates. Appl Geochem
63. Chen W, Chang AC, Wu L (2007) Assessing long-term environmental risks of trace elements in phosphate fertilizers. Ecotoxicol Environ Saf
64. Fang Y, Sun X, Yang W, Ma N, Xin Z, Fu J, Liu X, Liu M, Mariga AM, Zhu X et al (2014) Concentrations and health risks of lead, cadmium, arsenic, and mercury in rice and edible mushrooms in China. Food Chem
65. Panda SK, Upadhyay RK, Nath S (2010) Arsenic stress in plants. J Agron Crop Sci
66. Vodyanitskii YN (2013) Contamination of soils with heavy metals and metalloids and its ecological hazard (analytic review). Eurasian Soil Sci
67. Woolson EA, Axley JH, Kearney PC (1971) The chemistry and phytotoxicity of arsenic in soils: I. Contaminated field soils. Soil Sci Soc Am J
68. Smil V (2000) Phosphorus in the environment: natural flows and human interferences. Annu Rev Energy Environ

69. Chen M, Graedel TE (2016) A half-century of global phosphorus flows, stocks, production, consumption, recycling, and environmental impacts. Glob Environ Change
70. Van Kauwenbergh SJ (2010) World phosphate rock reserves and resources. IFDC-T-75
71. Abouzeid AZM (2008) Physical and thermal treatment of phosphate ores—an overview. Int J Miner Process
72. Scholz RW, Wellmer FW (2013) Approaching a dynamic view on the availability of mineral resources: what we may learn from the case of phosphorus? Glob Environ Change
73. Vaccari DA (2009) Phosphorus: a looming crisis. Sci Am
74. Elser J, Bennett E (2011) Phosphorus cycle: a broken biogeochemical cycle. Nature
75. Mew MC (2016) Phosphate rock costs, prices and resources interaction. Sci Total Environ
76. Suh S, Yee S (2011) Phosphorus use-efficiency of agriculture and food system in the US. Chemosphere
77. Cordell D, White S (2014) Life's bottleneck: sustaining the world's phosphorus for a food secure future. Annu Rev Environ Resour
78. Koppelaar RHEM, Weikard HP (2013) Assessing phosphate rock depletion and phosphorus recycling options. Glob Environ Change
79. Vogel C, Radtke M, Reinholz U, Schäfers F, Adam C (2015) Chemical state of chromium, sulfur, and iron in sewage sludge ash based phosphorus fertilizers. ACS Sustain Chem Eng
80. Kahiluoto H, Kuisma M, Ketoja E, Salo T, Heikkinen J (2015) Phosphorus in manure and sewage sludge more recyclable than in soluble inorganic fertilizer. Environ Sci Technol
81. Hanserud OS, Brod E, Øgaard AF, Müller DB, Brattebø H (2016) A multi-regional soil phosphorus balance for exploring secondary fertilizer potential: the case of Norway. Nutr Cycl Agroecosystems
82. Filippelli GM (2011) Phosphate rock formation and marine phosphorus geochemistry: the deep time perspective. Chemosphere
83. Mar SS, Okazaki M (2012) Investigation of Cd contents in several phosphate rocks used for the production of fertilizer. Microchem J
84. Cheraghi M, Lorestani B, Merrikhpour H (2012) Investigation of the effects of phosphate fertilizer application on the heavy metal content in agricultural soils with different cultivation patterns. Biol Trace Elem Res
85. Saueia CH, Mazzilli BP, Fávaro DIT (2005) Natural radioactivity in phosphate rock, phosphogypsum and phosphate fertilizers in Brazil. J Radioanal Nucl Chem
86. de Saussure T (2011) Recherches chimiques sur la végétation
87. Achard FK (1786) Chemische Untersuchung des Torfs. Crell's Chem Ann 2:391–403
88. Waksman SA (1932) Principles of soil microbiology. Williams & Wilkins, Baltimore
89. Flaig W (1966) The chemistry of humic substances. In: The use of isotopes in soil organic matter studies. Report of the FAO/IAEA Technical Meetings, Pergamon, New York, pp 103–127
90. Maillard LC (1912) Action des acides amines sur les sucres: Formation des melanoidines per voie methodologique. C R Acad Sci 154:66–68
91. Piccolo A (2001) The supramolecular structure of humic substances. Soil Sci
92. Piccolo A, Spaccini R, Savy D, Drosos M, Cozzolino V (2019) The soil humeome: chemical structure, functions and technological perspectives. In: Sustainable agrochemistry
93. Shindo H, Huang PM (1982) Role of Mn(IV) oxide in abiotic formation of humic substances in the environment. Nature
94. Frimmel FH (2000) Development in aquatic humic chemistry. Agronomie
95. Piccolo A (2016) In memoriam Prof. F.J. Stevenson and the question of humic substances in soil. Chem Biol Technol Agric
96. Zhang J, Wang J, An T, Wei D, Chi F, Zhou B (2017) Effects of long-term fertilization on soil humic acid composition and structure in Black Soil. PLoS One
97. Olk DC, Bloom PR, Perdue EM, McKnight DM, Chen Y, Farenhorst A, Senesi N, Chin Y-P, Schmitt-Kopplin P, Hertkorn N et al (2019) Environmental and agricultural relevance of humic fractions extracted by alkali from soils and natural waters. J Environ Qual

98. Esfahani MR, Stretz HA, Wells MJM (2015) Abiotic reversible self-assembly of fulvic and humic acid aggregates in low electrolytic conductivity solutions by dynamic light scattering and zeta potential investigation. Sci Total Environ
99. Baigorri R, Fuentes M, González-Gaitano G, García-Mina JM, Almendros G, González-Vila FJ (2009) Complementary multianalytical approach to study the distinctive structural features of the main humic fractions in solution: gray humic acid, brown humic acid, and fulvic acid. J Agric Food Chem
100. Drosos M, Jerzykiewicz M, Deligiannakis Y (2009) H-binding groups in lignite vs. soil humic acids: NICA-Donnan and spectroscopic parameters. J Colloid Interface Sci
101. Drosos M, Leenheer JA, Avgeropoulos A, Deligiannakis Y (2014) H-binding of size- and polarity-fractionated soil and lignite humic acids after removal of metal and ash components. Environ Sci Pollut Res
102. Weber J, Chen Y, Jamroz E, Miano T (2018) Preface: humic substances in the environment. J Soils Sediments
103. Gerke J (2018) Concepts and misconceptions of humic substances as the stable part of soil organic matter: a review. Agronomy
104. Rose MT, Patti AF, Little KR, Brown AL, Jackson WR, Cavagnaro TR (2014) A meta-analysis and review of plant-growth response to humic substances: practical implications for agriculture. In: Advances in agronomy
105. Canellas LP, Olivares FL, Aguiar NO, Jones DL, Nebbioso A, Mazzei P, Piccolo A (2015) Humic and fulvic acids as biostimulants in horticulture. Sci Hortic (Amsterdam)
106. Canellas LP, Olivares FL (2014) Physiological responses to humic substances as plant growth promoter. Chem Biol Technol Agric
107. Zanin L, Tomasi N, Cesco S, Varanini Z, Pinton R (2019) Humic substances contribute to plant iron nutrition acting as chelators and biostimulants. Front Plant Sci
108. Wu M, Wei S, Liu J, Liu M, Jiang C, Li Z (2019) Long-term mineral fertilization in paddy soil alters the chemical structures and decreases the fungistatic activities of humic acids. Eur J Soil Sci
109. Olk DC, Dinnes DL, Rene Scoresby J, Callaway CR, Darlington JW (2018) Humic products in agriculture: potential benefits and research challenges—a review. J Soils Sediments
110. Karpukhina E, Mikheev I, Perminova I, Volkov D, Proskurnin M (2019) Rapid quantification of humic components in concentrated humate fertilizer solutions by FTIR spectroscopy. J Soils Sediments
111. Pukalchik M, Kydralieva K, Yakimenko O, Fedoseeva E, Terekhova V (2019) Outlining the potential role of humic products in modifying biological properties of the soil—a review. Front Environ Sci
112. Shah ZH, Rehman HM, Akhtar T, Alsamadany H, Hamooh BT, Mujtaba T, Daur I, Al Zahrani Y, Alzahrani HAS, Ali S et al (2018) Humic substances: determining potential molecular regulatory processes in plants. Front Plant Sci
113. Leenheer JA (2009) Systematic approaches to comprehensive analyses of natural organic matter. Ann Environ Sci
114. Stevenson FJ (1994) Structural basis of humic substances. In: Humus chemistry: genesis, composition, reactions. ISBN 0471594741
115. John MK, Sprout PN, Kelley C (1965) The distribution of organic phosphorus in British Columbia soils and its relationship to soil characteristics. Can J Soil Sci 87–95
116. Elliott ET (1986) Aggregate structure and carbon, nitrogen, and phosphorus in native and cultivated soils. Soil Sci Soc Am J
117. Goh KM, Williams MR (1982) Distribution of carbon, nitrogen, phosphorus, sulphur, and acidity in two molecular weight fractions of organic matter in soil chronosequences. J Soil Sci
118. Jensen MP, Nash KL, Morss LR, Appelman EH, Schmidt MA (1996) Immobilization of actinides in geomedia by phosphate precipitation. ACS Symp Ser
119. Macholz R (1987) Phytic acid—chemistry & applications. Herausgegeben von E. Graf. 344 Seiten, zahlr. Abb. und Tab. Pilatus Press Minneapolis 1986. Preis: 64,95 $. Food/Nahrung

120. Nanny MA, Minear RA (1997) Characterization of soluble unreactive phosphorus using ^{31}P nuclear magnetic resonance spectroscopy. Mar Geol
121. Newman RH, Tate KR (1980) Soil phosphorus characterisation by ^{31}P nuclear magnetic resonance. Commun Soil Sci Plant Anal
122. Makarov MI, Guggenberger G, Zech W, Alt HG (1996) Organic phosphorus species in humic acids of mountain soils along a toposequence in the Northern Caucasus. Z fur Pflanzenernahr und Bodenkd
123. Norman EC, Graham CL, Hayes MHB (1997) Investigations into the nature of phosphorus in soil humic acids using ^{31}P NMR spectroscopy. In: Humic substances, peats and sludges
124. Condron LM, Frossard E, Condron LM, Frossard E (1997) Use of ^{31}P NMR in the study of soils and the environment. In: Nuclear magnetic resonance spectroscopy in environment chemistry
125. Busato JG, Canellas LP, Rumjanek VM, Velloso ACX (2005) Phosphorus in an inceptsoil under long-term sugarcane: II—humic acid analysis by NMR^{31}P. Rev Bras Cienc do Solo
126. Savy D, Mazzei P, Nebbioso A, Drosos M, Nuzzo A, Cozzolino V, Spaccini R, Piccolo A (2016) Molecular properties and functions of humic substances and humic-like substances (hulis) from biomass and their transformation products. In: Analytical techniques and methods for biomass. ISBN 9783319414140
127. He Z, Olk DC, Cade-Menun BJ (2011) Forms and lability of phosphorus in humic acid fractions of Hord silt loam soil. Soil Sci Soc Am J
128. Li M, Mazzei P, Cozzolino V, Monda H, Hu Z, Piccolo A (2015) Optimized procedure for the determination of P species in soil by liquid-state ^{31}P-NMR spectroscopy. Chem Biol Technol Agric
129. Bartoszek M, Polak J, Sułkowski WW (2008) NMR study of the humification process during sewage sludge treatment. Chemosphere
130. Sposito G (2016) The chemistry of soils
131. Zamparas M, Gianni A, Stathi P, Deligiannakis Y, Zacharias I (2012) Removal of phosphate from natural waters using innovative modified bentonites. Appl Clay Sci
132. Zamparas M, Drosos M, Georgiou Y, Deligiannakis Y, Zacharias I (2013) A novel bentonite-humic acid composite material Bephos^{TM} for removal of phosphate and ammonium from eutrophic waters. Chem Eng J 225:43–51
133. Zamparas M, Drosos M, Georgiou Y, Deligiannakis Y, Zacharias I (2013) A novel bentonite-humic acid composite material Bephos^{TM} for removal of phosphate and ammonium from eutrophic waters. Chem Eng J
134. Zamparas M, Drosos M, Deligiannakis Y, Zacharias I (2015) Eutrophication control using a novel bentonite humic-acid composite material Bephos^{TM}. J Environ Chem Eng 3:3030–3036
135. Zamparas M, Kapsalis VC, Kanteraki AE, Vardoulakis E, Kyriakopoulos GL, Drosos M, Kalavrouziotis IK (2019) Novel composite materials as P-adsorption agents and their potential application as fertilizers. Glob Nest J
136. Zamparas M, Drosos M, Deligiannakis Y, Zacharias I (2015) Eutrophication control using a novel bentonite humic-acid composite material Bephos^{TM}. J Environ Chem Eng
137. Cozzolino V, Di Meo V, Piccolo A (2013) Impact of arbuscular mycorrhizal fungi applications on maize production and soil phosphorus availability. J Geochem Explor
138. Zamparas M (2021) The role of resource recovery technologies in reducing the demand of fossil fuels and conventional fossil-based mineral fertilizers. In: Low carbon energy technologies in sustainable energy systems. Elsevier, Amsterdam, pp 3–24
139. Behrens S, Appel I (2016) Magnetic nanocomposites. Curr Opin Biotechnol 39:89–96
140. Zamparas M, Kyriakopoulos GL, Drosos M, Kapsalis VC, Kalavrouziotis IK (2020) Novel composite materials for lake restoration: a new approach impacting on ecology and circular economy. Sustainability
141. Ramasahayam SK, Gunawan G, Finlay C, Viswanathan T (2012) Renewable resource-based magnetic nanocomposites for removal and recovery of phosphorous from contaminated waters. Water Air Soil Pollut 223:4853–4863

Chapter 8
A New Method for Lake Restoration, Impacting on Circular Economy (CE)

Miltiadis G. Zamparas and Grigorios L. Kyriakopoulos

Abstract

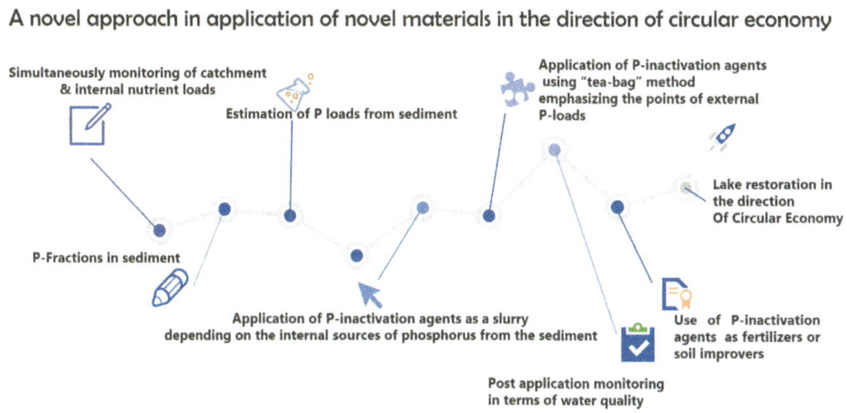

Graphical Abstract of the Chapter

The key aim of the chapter is to incorporate innovative technology that can be used to treat eutrophic waters. A new environmentally friendly method is focused on the "teabag" process, under which P-inactivation agents were stored in water-permeable bags and submerged in the water column to trap P, which is one of the primary causes of eutrophication. In specific, composite lanthanum-modified bentonite (LMB) and Fe-modified bentonite (FMB) were investigated, and bench-scale batch tests were performed to examine their adsorption efficacy as slurry and as tea-bag. Moreover, the adsorption isotherms and kinetics were also studied. Experimental outcomes revealed

M. G. Zamparas (✉)
Laboratory of Sustainable Waste Technology Management, School of Science and Technology, Hellenic Open University, Parodos Aristotelous 18, 26335 Patras, Greece
e-mail: mzamparas@upatras.gr

G. L. Kyriakopoulos
Photometry Laboratory, Electric Power Division, School of Electrical and Computer Engineering, National Technical University of Athens, 15780 Athens, Greece

M. G. Zamparas and G. L. Kyriakopoulos (eds.), *Chemical Lake Restoration*,
https://doi.org/10.1007/978-3-030-76380-0_8

that LMB and FMB (as teabag) achieved a maximal sorption capacity of 7.80 and 25.1 mg/g, accordingly, being considered as acceptable P rates for natural eutrophic waters. The "teabag" approach avoids the material covering the bottom sediment, thereby eliminating the occurrence of smothering of benthos. Also, P-adsorbents are also suitable for other agricultural uses like that of soil improver. Conclusively, it was signified the of utmost importance perspective of LMB and FMB composite clays' recycling at later agricultural uses, since it sustains beneficial ecological impacts: routes of the circular economy (CE).

Keywords Lake restoration · Eutrophication management · Tea-bag · Circular economy

8.1 Introduction

Nowadays, the growing world population, the increased food demands and the intensification of agricultural development are all closely correlated with one another. Although seafood does only reach a nutrition shortage to humans, the sharp increase of global population is deterministically causing agricultural intensification, the extensive usage of fertilizers, and it is, thereupon, triggering eutrophication incidences. The contamination of aqueous ecosystems is a significant trigger of eutrophication, while formulating toxic algal blooms, a reduction of dissolved oxygen to water bodies, which jeopardizes the local ecology [1–3], on the basis of social, economics and technology transitions reported [4, 5]. The remediation of degraded eutrophic waters has focused on reducing the loading of nutrients from the catchment and runoff from farm soils [6–8]. Besides, at eutrophic conditions, the degraded waters' recovery from extreme sediment P loads are prone to time delays. Under eutrophic circumstances, elevated amounts of phosphorus persist in the water, which hinders the regeneration of the ecosystem even though decreased external loading [9].

Bephos™ (FMB) is an innovative low-cost engineered material consisting of Fe, Cu and humic acid (HA) that have been embedded in the matrix of natural bentonite [10, 11]. Furthermore, Phoslock™ (or LMB) is a P inactivation technology, where lanthanum ions in the structure of bentonite clay can adsorb soluble reactive phosphorus (SRP) [12, 13]. Today, Phoslock™ is being gradually used to adsorb phosphorus (mainly in the inorganic form of phosphates) from the water column while blocking P flux from bottom sediment that applies like an active barrier system (ABS) [14–17].

Regarding the monitoring of nutrients at lake sediment, it is noteworthy the development of a vertical flow of sediments by bioturbating with time. This vertical flow would replenish the sediment's phosphorus level, and will further remediate the contaminated sediment. The efficacy of regulating all of the P-total and N-soluble can be measured by aeration, while bioturbation harms nitrogen elimination, and has little on phosphorus control performance [18].

Nowadays, major research studies into the fate of nutrients arise in wetlands and reservoirs [19]. The primary role of LMB application to these water bodies is that it is an appropriate agent for the elimination of soluble phosphorus, which affects algae bloom [20, 21]. In specific, Bishop and Richardson [12, 19], firstly, reported P and N spikes after processing the water, following the implementation of LMB to irrigation ponds and, secondly, monitored the post-treatment of nutrients' loading and blooms of chronic cyanobacteria. These authors stressed the necessity to assess the mitigation of P sediment at nutrient ratios' shift and modification of the composite profile of algal assemblage [19]. Moreover, these authors showed the crucial role of finding the suitable technology of P binding for the regulation of algae assemblage composition and water quality parameters. Moreover, in an earlier analysis, Robb et al. [17] conducted laboratory trials and reported the LMB efficiency to attach sediment bioavailable P using less than a millimeter thickness of the clay, particularly at extreme fluvial blooms of blue-green algae in the summer months.

LMB has been comprehensively examined in the relevant literature [22–24] while another group of studies unveiled that the application of LMB concentration causes conductivity increase and turbidity decrease due to quick settlement [24, 25]. Moreover, LMB can be successfully applied to control sediment P-flux under anaerobic conditions, while pre- and post-treatment of LMB included the parameters of pH, total suspended solids, conductivity, turbidity, color and phosphate concentration; accomplishing 95% removal of phosphate anions PO_4^{3-}, while LMB application did not cause pH change at the final effluent [22], while such pre- and post-treatment of LMB showed that the pH of the treated water returns closer to the neutral level within days of LMB application [23]. Moreover, LMB can bind soluble reactive P and to release nutrients showing NH_3-release of small quantities, while no affecting oxygen saturation [24] and pH [25]. Multiple smaller doses of LMB, comparing to a single high dose, can release sedimentary P under aerobic conditions comparing to significant P release occurring with persistent and prolonged anaerobic conditions [26]. At a laboratory-tested analysis, LMP revealed that monovalent phosphate anion of $H_2PO_4^-$ showed high affinity for the adsorbent surface, while activation energy was lower at a higher pH since the loss of adsorption sites at higher pH is unaffected by ionic strength [27].

The key challenges are focused on selecting innovative materials in proper conditioning, which can be introduced as "teabag" into the water column thus, absorbing P from the water column and protecting endangered ecosystems. The necessary quantities of material utilized in the form of slurry should be accurately calculated based on the potentially bioavailable mobile P of the sediment, and its external flows can be adsorbed by the proposed method of "teabag". Other challenging issues are related to improvements of P use in efficient and sustainable-to-environment ways. In this research context of removing P-containing compounds at water bodies, the proposed technologies should use functional sorbents regarding surface properties for capture and removal. Among such technologies, the application of LMB and FMB materials and their recycling capabilities as alternative soil fertilizers by implementing the teabag method are valued as among the most promising ones.

8.2 Materials and Methods

8.2.1 Materials

The anthropogenic-origin material produced by the Land and Water Division of Australia's CSIRO, was lanthanum bentonite (LMB). The chapter reports the synthesis of Fe-modified bentonite, which has been documented in Zamparas et al. [12].

8.2.2 Batch Experiments

Laboratory measurements of phosphate sorption were described by Zamparas et al. [28]. The calculation of adsorption isotherms was rendered by batch equilibration of 0.02 g of the bentonite sample with 50 mL of aqueous nutrient solutions. Initial P-concentrations range was 0.05–5 mg/L. The tests were performed at room temperature (25 ± 1 °C) and pH 7 for 3 h. After equilibration, the sample was then subjected to centrifugation (4500 rpm for 15 min) and the molybdate blue spectrophotometric process was then utilized to calculate the concentration of the supernatant (ce) [29]. Sorption kinetic results of phosphates on two modified clays at contract times varying between 15 and 240 min were analyzed. The optimum conditions at the experimental scale of analysis were that of pH 7, P concentration 0.1 mg/L, adsorbent dose 0.02 g and 25 ± 1 °C temperature.

8.3 Results and Discussion

8.3.1 Adsorption Isotherms

Langmuir and Freundlich equations were introduced to estimate sorption isotherms of PO_4–P onto the LMB and FMB, and to calculate the maximum of PO_4–P sorbed on each material (as slurry and as tea-bag). The rates of PO_4–P adsorption on LMB and FMB as a result of the process are shown in Fig. 8.1a and b, respectively. Furthermore, the maximum phosphate adsorption capacity of the LMB in the form of slurry was 11.28 mg/g, and in the form of "tea-bag" was 7.80 mg/g.

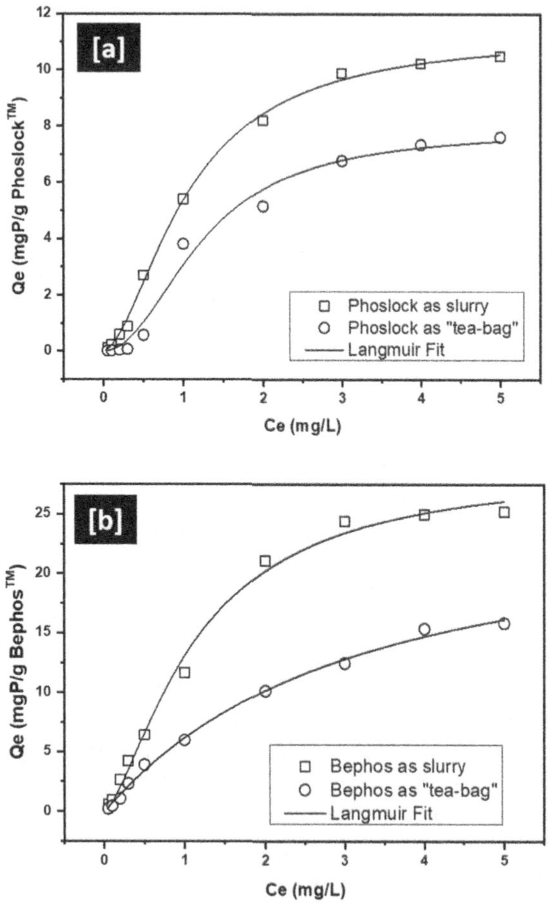

Fig. 8.1 **a** Adsorption isotherms of phosphate by lanthanum modified bentonite (LMB) as a slurry and as "teabag". **b** Adsorption isotherms of phosphate by Fe-modified bentonite (FMB) as a slurry and as "teabag". P concentration 0.05–5 mg/L, adsorbent dose 0.02 g, pH = 7.0, T = 25 °C. *Source* Authors' own study

8.3.2 Adsorption Kinetics

The adsorption of different compounds in tea bags is not as appears, because the tea bags are functioning as a barrier that physically slows down the kinetics of the binding, moving the equilibrium period to more than three hours. In the case of FMB, a pause is more prominent owing to dimensional variations between FMB and lamellar.

Adsorption kinetic data of PO_4–P on the modified innovative clays LMB and FMB are also examined (data not shown). The plots represent the adsorption of phosphates onto LMB and FMB versus time extending from 15 to 240 min, as a function of the method of introducing the material into the aqueous solution (as teabag and as slurry). The findings reveal that most of PO_4–P were adsorbed during the first 60 min of the experiment, then, sorption equilibrium commences to establish. It is crucial to suggest that the adsorption rate of P onto both modified clays did not vary depending on the implementation method into the aqueous solution regardless, seemingly, the adsorption capacity of the materials in tea bags is reduced. In particular, tea bags are operating as a barrier that is physically slowing down the binding kinetics of the binding, while delaying the equilibrium period up to more than three hours. FMB delay is even more pronounced, mainly due to different structures of FMB and LMB.

8.4 Challenging Issues of Novel Composite Materials in the Direction of Ecology and Circular Economy (CE)

In general, it has been stated that the usage of innovative modified materials for chemical lake restoration can minimize the internal P flux from lake bottom regardless that the physicochemical properties of P adsorbents are largely decided by a broad range of environmental factors involved to distinct control and risks' research, thereby necessitating thorough investigation of each one material before implementation [30].

Thus, the current study was concentrated on the efficiency and sustainability of using phosphorus as an important resource in the environment. In this context, tools for the removal of P-containing compounds from water using innovative sorbents should primarily regard surface properties for capture and removal, by recycling the LMB and FMB materials as unconventional soil improvers or fertilizers.

Smothering of benthos and accumulation of sediments—is related to, first, the deposition of layers of sediment on the seabed and, secondly, the depth of vertical overburden (mainly associated with the trophic group and particle size/sediment type [31],—can be directly affected by both the sea disposal of dredged materials and by the application of capping materials. Because a majority of benthic species reside in the top 10 cm from the seabed, they must retain some relation to the sediment–water interface due to feeding needs [32].

Many aquatic organisms are adapted to re-surface from thin deposits, however, the extreme deposition events are deemed a significant burden on the environment. Consequently, the implementation action of P sorbents in a natural and dynamic bionetwork necessitates careful consideration. The required amounts of material employed in the form of slurry should be precisely assessed, being based on the bioavailable P fractions of the sediment [30, 33–35] present in the water column and its external inflows can be adsorbed by the recommended approach of "tea-bag".

"Tea-bag" can, firstly, introduce the innovative clays in the water column and, secondly, support a significant reduction of the, otherwise, unnecessary amount of the P-inactivation agents disposed in degraded water bodies. In practice, after monitoring the ecosystem (internal P-loads, bathymetry, etc.) the "teabag" method can adsorb the bioavailable P from the water. P-inactivation agents can be connected to buoys at various depths to absorb the available inorganic P remaining in the water without environmental risking. Modified clays per bag are applied at varying quantities related to P concentrations in the water column, the spatial sites of major P inputs and the total volume of the lake. To this end it is utmost importance the capability of LMB and

FMB recycling and reuse for farming applications. These restorative clay-based tools can be added to the soil without interrupting the soil microenvironment but increasing its cation exchange capacity (CEC) [33]. Furthermore, the elemental composition of FMB in Fe and Cu is supporting its micronutrient properties [11, 12].

8.5 Conclusions

According to the findings above, the key results mentioned are evidenced in Table 8.1. Based on Table 8.1 the study outcomes were analyzed by utilizing weighted indicators. The suggested approaches are primarily useful from an environmental viewpoint; accompanied by a promising perspective in the sense of the resources used and economic efficiency. The explanation for this modest materials' and economics' performance resides in the reality of having no substantial prior experience of valuable and widespread use. However, considering the combined assessment of LMB and FMB, it is possible to monitor the volumes of slurry eventually disposed to eutrophic water bodies with greater precision. The amounts of material utilized in the form of slurry are accurately calculated, based on potentially bioavailable P from sediments. Inorganic phosphorus presented in the water column and its external flows can be adsorbed by the suggested "tea bag" method.

Such an applied methodology can accomplish great outcomes in terms of reducing the slurry amount of innovative materials. The environmental remediation of eutrophic environments was achieved by the sorption of P to accurately estimated amounts of modified Fe-bentonite and modified La-bentonite, where the generated "teabag" product can be additional used as a soil improver or fertilizer. Moreover, management of natural mineral resources—such as rock phosphate RP—is accomplished, since P-fertilizers can be generated by innovative modified clays. The simplicity, availability and low cost of all the materials applied, as well as the high P removal levels achieved, ensure the practicality of the proposed method.

Table 8.1 Evaluation profile of the method developed. *Source* Authors' own study

Key-aspects of future development	Material	Environment	Economy
Technological maturity/industrial scalability	++	+	+
Technical adaptability to other techniques of environmental remediation	+++	+++	++
Ecosystems' protection and sustainability	+	+++	++
Economic scalability	+	+++	++

Scale of weighted indicators: +: low positive, ++: fairly positive, +++: highly positive

References

1. Lin J, Zhan Y, Zhu Z (2011) Evaluation of sediment capping with active barrier systems (ABS) using calcite/zeolite mixtures to simultaneously manage phosphorus and ammonium release. Sci Total Environ
2. Nixon SW (1995) Coastal marine eutrophication: a definition, social causes, and future concerns. Ophelia
3. Solheim AL, Rekolainen S, Moe SJ, Carvalho L, Phillips G, Ptacnik R, Penning WE, Toth LG, O'Toole C, Schartau A-KL et al (2008) Ecological threshold responses in European Lakes and their applicability for the Water Framework Directive (WFD) implementation: synthesis of lakes results from the REBECCA project
4. Kapsalis VC, Kyriakopoulos GL, Aravossis KG (2019) Investigation of ecosystem services and circular economy interactions under an inter-organizational framework. Energies
5. Kyriakopoulos GL, Kapsalis VC, Aravossis KG, Zamparas M, Mitsikas A (2019) Evaluating circular economy under a multi-parametric approach: a technological review. Sustainability 11
6. Hickey CW, Gibbs MM (2009) Lake sediment phosphorus release management—decision support and risk assessment framework. N Z J Mar Freshw Res 43:819–856
7. Zamparas M, Gianni A, Stathi P, Deligiannakis Y, Zacharias I (2012) Removal of phosphate from natural waters using innovative modified bentonites. Appl Clay Sci 62–63:101–106
8. Zamparas M, Zacharias I (2014) Restoration of eutrophic freshwater by managing internal nutrient loads. A review. Sci Total Environ 496:551–562
9. Jeppesen E, Sondergaard M, Jensen JP, Havens KE, Anneville O, Carvalho L, Coveney MF, Deneke R, Dokulil MT, Foy B et al (2005) Lake responses to reduced nutrient loading—an analysis of contemporary long-term data from 35 case studies. Freshw Biol 50:1747–1771
10. Zamparas M, Drosos M, Deligiannakis Y, Zacharias I (2015) Eutrophication control using a novel bentonite humic-acid composite material Bephos^{TM}. J Environ Chem Eng 3:3030–3036
11. Zamparas M, Drosos M, Georgiou Y, Deligiannakis Y, Zacharias I (2013) A novel bentonite-humic acid composite material Bephos^{TM} for removal of phosphate and ammonium from eutrophic waters. Chem Eng J
12. Bishop WM, McNabb T, Cormican I, Willis BE, Hyde S (2014) Operational evaluation of Phoslock phosphorus locking technology in Laguna Niguel Lake, California. Water Air Soil Pollut
13. Kasprzyk M, Obarska-Pempkowiak H, Masi F, Gajewska M (2018) Possibilities of Phoslock® application to remove phosphorus compounds from wastewater treated in hybrid wetlands. Ecol Eng
14. Spears BM, Lürling M, Yasseri S, Castro-Castellon AT, Gibbs M, Meis S, McDonald C, McIntosh J, Sleep D, Van Oosterhout F (2013) Lake responses following lanthanum-modified bentonite clay (Phoslock®) application: an analysis of water column lanthanum data from 16 case study lakes. Water Res
15. Haghseresht F, Wang S, Do DD (2009) A novel lanthanum-modified bentonite, Phoslock, for phosphate removal from wastewaters. Appl Clay Sci
16. Liu Y, Tian K, Winks A (2009) Water eutrophication prevention by phoslock flocculation and sediment capping reaction. In: Proceedings of the 2009 international conference on environmental science and information application technology, ESIAT 2009
17. Robb M, Greenop B, Goss Z, Douglas G, Adeney J (2003) Application of Phoslock^{TM}, an innovative phosphorus binding clay, to two Western Australian waterways: preliminary findings. Hydrobiologia
18. Yin H, Zhu J, Tang W (2018) Management of nitrogen and phosphorus internal loading from polluted river sediment using Phoslock® and modified zeolite with intensive tubificid oligochaetes bioturbation. Chem Eng J
19. Bishop WM, Richardson RJ (2018) Influence of Phoslock® on legacy phosphorus, nutrient ratios, and algal assemblage composition in hypereutrophic water resources. Environ Sci Pollut Res

20. Liu Y, Tian K, Winks A (2009) Water eutrophication prevention by phoslock flocculation and sediment capping reaction. In: Proceedings of the 2009 international conference on environmental science and information application technology, ESIAT 2009, IEEE, vol 1, pp 349–352
21. Liu Y, Tian K, Winks A (2009) Phoslock, a soluble phosphorus binding technology, application in Lake Dianchi eutrophic water purification. In: Proceedings of the 2009 international conference on environmental science and information application technology, ESIAT 2009
22. Kasprzyk M, Gajewska M (2017) Preliminary results from application Phoslock® to remove phosphorus compounds from wastewater. J Ecol Eng
23. Liu Y, Tian K, Winks A (2009) Phoslock®: an effective technology for effective dissolved phosphorus removal in extensive water bodies under a wide range of chemical conditions. In: Proceedings of the 3rd international conference on bioinformatics and biomedical engineering, iCBBE 2009
24. van Oosterhout F, Lürling M (2013) The effect of phosphorus binding clay (Phoslock®) in mitigating cyanobacterial nuisance: a laboratory study on the effects on water quality variables and plankton. Hydrobiologia
25. Yamada-Ferraz TM, Sueitt APE, Oliveira AF, Botta CMR, Fadini PS, Nascimento MRL, Faria BM, Mozeto AA (2015) Assessment of Phoslock® application in a tropical eutrophic reservoir: an integrated evaluation from laboratory to field experiments. Environ Technol Innov
26. Meis S, Spears BM, Maberly SC, Perkins RG (2013) Assessing the mode of action of Phoslock® in the control of phosphorus release from the bed sediments in a shallow lake (Loch Flemington, UK). Water Res
27. Haghseresht F, Wang S, Do DD (2009) A novel lanthanum-modified bentonite, Phoslock, for phosphate removal from wastewaters. Appl Clay Sci 46:369–375
28. Zamparas M, Deligiannakis Y, Zacharias I (2013) Phosphate adsorption from natural waters and evaluation of sediment capping using modified clays. Desalin Water Treat
29. Gianni A, Zamparas M, Papadas IT, Kehayias G, Deligiannakis Y, Zacharias I (2013) Monitoring and modeling of metal concentration distributions in anoxic basins: Aitoliko Lagoon, Greece. Aquat Geochem 19:77–95
30. Zamparas M, Drosos M, Kalavrouziotis IK (2019) Sorption of phosphate from innovative composite material focusing on physicochemical interactions. Desalin Water Treat 151:212–218
31. Miller DC, Muir CL, Hauser OA (2002) Detrimental effects of sedimentation on marine benthos: what can be learned from natural processes and rates? Ecol Eng 19:211–232
32. Bolam SG, Barrio-Frojan CRS, Eggleton JD (2010) Macrofaunal production along the UK continental shelf. J Sea Res 64:166–179
33. Zamparas M, Kyriakopoulos GL, Kapsalis VC, Drosos M, Kalavrouziotis IK (2019) Application of novel composite materials as sediment capping agents: column experiments and modelling. Desalin Water Treat 170:111–118
34. Zamparas M, Gavriil G, Coutelieris FA, Zacharias I (2015) A theoretical and experimental study on the P-adsorption capacity of Phoslock™. Appl Surf Sci 335:147–152
35. Zamparas M, Kapsalis V, Kanteraki AE, Vardoulakis E, Kyriakopoulos GL, Drosos M, Kalavrouziotis IK (2019) Novel composite materials as P-adsorption agents and their potential application as fertilizers. Glob Nest J 21:48–57

Index

M. G. Zamparas and G. L. Kyriakopoulos (eds.), *Chemical Lake Restoration*,
https://doi.org/10.1007/978-3-030-76380-0

The manufacturer's authorised representative in the EU is Springer Nature Customer Service Centre GmbH, Europaplatz 3, 69115 Heidelberg, Germany. If you have any concerns regarding our products, please contact ProductSafety@springernature.com

Printed and bound by CPI Group (UK) Ltd, Croydon, CR0 4YY
28/09/2026
02224091-0003